DIAMOND DEALERS
AND
FEATHER MERCHANTS

DIAMOND DEALERS
DEALERS
AND
FEATHER
MERCHANTS

Tales from the Sciences

Irving M. Klotz

BIRKHÄUSER
Boston • Basel • Stuttgart

Library of Congress Cataloging-in-Publication Data

Klotz, Irving M. (Irving Myron), 1916-
  Diamond dealers and feather merchants.
1. Science — Miscellanea. I. Title.
Q173.K5735 1985   500   85-7549
ISBN 0-8176-3303-0

CIP — Kurztitelaufnahme der Deutschen Bibliothek

Klotz, Irving M.:
Diamond dealers and feather merchants: tales from the sciences / Irving M. Klotz. — Boston;
Basel; Stuttgart: Birkhäuser, 1985.
  ISBN 3-7643-3303-0 (Basel ...)
  ISBN 0-8176-3303-0 (Boston)

© Birkhäuser Boston, 1986

9   8   7   6   5   4   3   2   1

ISBN 0-8176-3303-0
ISBN 3-7643-3303-0

Manufactured in the United States of America

To the Memories of
Arthur C. Lunn
(19 February 1877–19 November 1949)
and
Julius Stieglitz
(26 May 1867–10 January 1937)
Professors at the University of Chicago
who opened windows, gently.

Contents

Preface

To paraphrase Saul Bellow, it is extremely difficult to escape from the conceptual bottles into which we have been processed, or even to become aware that we are confined within them. Anthropocentrism, an ancient tradition, is an intellectual constraint that has continually impeded objective probing of the universe around and within us. We are probably born with that constriction, perhaps as a result of evolutionary selection or because each of us has been created in the image of the Deity. But it is only the core of our mental "gestalt." Around it we find additional shells of intellectual obstructions deposited by accretion from our family, our teachers, our experiences and the society in which we are immersed.

It is very hazardous to embrace novel scientific ideas. Personal and social experiences show that the vast majority turn out to be failures. What standards can one use to make judgments? There is a universal tendency to rely on "common sense;" but as Einstein pointed out, this is a collection of views, sensible or not, imprinted in us before the age of sixteen. I have found it a challenge to convince young students that much of what they are certain about and, in fact, correct about, is actually contrary to common sense. For example, on any bright day, anyone who is not blind or an idiot can see the sun literally moving around the earth, from east to west. Why does any lay person of modern times unthinkingly accept the notion that it is we on earth who are "really" moving? He or she would certainly not do so in following the motion of an airplane in the sky. Anthropocentrism and religious beliefs were fortified by common sense to justify rejection of Copernicanism. The shock of heliocentrism arose from the conflict with all of these principles.

Analogous comments can be made about evolution or relativity theory, but in these circumstances a complementary facet of common sense must be noticed. Our common experience is very limited. We have no personal experiences at very high speeds (near that of light), of very long times (many millennia), of very tiny objects (atoms), or of very great distances (to cosmic black holes). We may fool ourselves into thinking that the extraordinary phenomena encountered in these areas are analogous to some common experiences because someone uses similar words to describe them. But words have very different meanings in a scientific context, and superficial transpositions to common experience may rob them of any sense, even in sentences that obey all the rules of grammar and rhetoric.

Sometimes, a sounder version of common sense is used as a basis of judgment in the sciences. Often a radically new idea can be shown to be in gross conflict with absolutely sound experimental or theoretical principles, i.e. unassailable principles at the foundation of the science of the time. In the overwhelming majority of such circumstances, the novel idea will disappear into the limbo of human errors. But, of course, what is often brought to public attention, post hoc, are the rare exceptions for which contradictions were ultimately resolved in favor of the novel, path-opening concept. It is easy to recognize after the event, usually a long time afterward, where the hidden, unwarranted premises of the traditional scientific doctrine led us astray. But before the event, as one of the great physicists of the twentieth century, Niels Bohr, remarked, "Prediction is very difficult, especially of the future."

Scientists are first of all people. Some are saints, a few are charlatans, most try to abide by unwritten codes of good scientific practice. There are a few leaders, many followers, occasional rebels. Some act like intellectual dinosaurs, others are implacable skeptics, a few are free-floating undisciplined mystics, most go along with accepted doctrines. There are diamond dealers, feather merchants and those struggling to earn an honest scientific living. At some stage in his or her scientific life, an individual may fit into each of these categories. All of these facets of temperament and personality color one's view of the intellectual landscape.

Although we rarely realize the extent to which we are prisoners of our own ideas, we can frequently recognize that *other* people are predisposed to error. What we cannot discern in ourselves becomes glaringly evident in others. The aim of this small volume is to provide a guided, non-random tour through a few of the galleries of science to view other people's fallibilities, to explore or hint at facets of temperament, personality and social, political, and religious environment that led them to make monumental misjudgments. This collection of vignettes may supply each of us with some tools to "release ourselves from the bottles into which *we* have been processed."

I

Bending Perception to Wish: The Future as Froth and Fantasy

Every living enterprise spawns new ideas. Like mutants of living organisms, most of these novelties are self-lethal. After a short or long exposure to the selection pressures of their environment, they vanish. A few may survive. Rare ones may even flourish and in time change the entire complexion of the enterprise.

To practitioners in any creative area, the attempt to recognize the ultimate viability and potential of a novel conception at its birth, rather than at its maturity, has always been a tantalizing challenge. In the natural sciences, this aim has been professed continuously. In principle in the sciences, we encourage creativity or novelty; in practice we have great difficulty recognizing it.

For those who have attained some stature and responsibility in their fields, the problem becomes particularly acute. In the course of professional advancement, many scientists become counselors to granting agencies of the federal government or of private foundations, members of editorial boards of research journals or managers of large industrial research enterprises. In these positions one is faced periodically with a presentation of a very novel, unorthodox or radical proposal containing suggestions or modes of thinking gratingly contrary to current, accepted paradigms. Is the author a crackpot or a genius? Most individuals in decision-making posts in science, in addition to making their own preliminary assessment, turn to external authorities in the field for advice and guidance. Suppose you were an editor and had received the following external assessment of a paper; what would you do?

> I have searched the tedious studies of this author without success for some trace of ingenuity, acuteness or learning that might compensate for his evident deficiency in the powers of solid thinking or of calm and patient investigation.... This manuscript teaches no new truth, reconciles no contradictions, arranges no anomalous facts, suggests no new experiments and leads to no new inquiries.... As this paper contains nothing which deserves the name either of experiment or discovery, and as it is in fact destitute of every species of merit, it should certainly be admitted to your Proceedings to join the company of that multitude of other paltry and unsubstantial papers which are being published in your journal.

This was the essence of a qualified reviewer's opinion (adapted here from the published review) of a paper that later turned out to be one of the landmarks of science. The reviewer, Henry Brougham, has long been forgotten; the author of the paper was Thomas Young, the nineteenth century pioneer of the wave nature of light.

On occasion, during tenures on grants panels and editorial boards I have been in a somewhat similar position, albeit, fortunately, on issues of lesser significance. I have wondered what precautions might predispose reviewers to be more flexible. Certainly the acquisition of some historical perspective should help. All the new young assistant professors I encounter think that the history of science in their field started the day they arrived on campus. I always tell them they are wrong — it started the day *I* arrived on campus.

For some decades I have searched the literature of science, during occasional spare time, to unearth some of the more outrageous blunders in judgment or perception made by distinguished and highly respected scientists of the past. I have found many examples that provide amusing and instructive insights. Since scientific writing was less constrained in the past and since many individuals developed hyperthyroid styles of expresssion and had the poor foresight to commit their opinions to print, we have an opportunity today to read some of these memorable outbursts.

For the first of these I must present a few prefatory comments.

In line with the wishes of Creationists, let me start by pointing out that God created optical activity [Genesis 1:3] 5,745 years ago (as of 1985). However, it was only about 180 years ago that man discovered optical activity. For nonscientist readers who may not know what optical activity is, let me interject that it is a property of many materials, such as Polaroid lenses, that endows them with the capability to flatten the multidirectional oscillations within a beam of light waves into vibrations in a single two-dimensional planar ribbon. Until about a hundred years ago the molecular basis for optical activity was a mystery. However, in 1875 an obscure twenty-three-year-old Dutch chemist, whose first job was as a faculty member of the agricultural college at Utrecht, published a thin theoretical book that was translated into German not long afterward with the title *The Arrangement of Atoms in Space.* Shortly after the publication of the German edition, there appeared a note in the *Journal für praktische Chemie* written by the editor, Adolph Wilhelm Hermann Kolbe, professor at the great German university of Leipzig. Kolbe was one of the most distinguished chemists of his time, famous (among many other achievements) for the first total synthesis (in 1845) of an organic compound (acetic acid) from its constituent elements. Let us examine a transcription of his remarks, freely translated.

Recently I pointed out that deficiences in fundamental chemical knowledge and the lack of a good liberal education characteristic of many professors is responsible for the current decline of chemical research. As a result of this deplorable situation, there has been a proliferation of that weed, seemingly erudite and profound but actually trivial and shallow natural philosophy. This style of explication, expunged fifty years ago by exact natural science, is now being retrieved, by scientific quacks, from the junk pile of man's errors. Flashily dressed and covered with a lot of makeup like a worn-out old hooker, she is smuggled into respectable society, where she does not belong. If anyone thinks my concerns are exaggerated, he should read, if he can stomach it, a recent monograph by a Mr. van't Hoff, entitled *The Arrangement of Atoms in Space,* a book swollen with infantile foolishness....This young punk employed by the Cow College at Utrecht, evidently has no taste for exact chemical investigation. He prefers to mount his winged horse (Pegasus), evidently borrowed from the barns of the Cow College, and to proclaim that upon his bold ascent to Mount Parnassus he had a vision of atoms arranged in space. The sober-minded chemical world does not relish such hallucinations.

It is characteristic of these days — uncritical and anti-intellectual — that a virtually unknown chemist from a veterinary school has the gall to make pronouncements on the most fundamental problems of chemistry — which may never be solved — and to propose solutions with a self-assurance and insolence that can only astonish true scientists.

These excerpts impart the flavor of Kolbe's attack, which appeared prominently in his editorial column "Signs of the Times." Such was

Kolbe's assessment of the contribution of young Jacobus Henricus van't Hoff (Figure 1), one of the most creative minds of nineteenth-century chemistry. After his work in organic chemistry leading to the grand concept of the asymmetric tetrahedral carbon atom, van't Hoff began his studies in chemical thermodynamics and made seminal contributions that place him among the primary founders of physical chemistry. In fact he and Ostwald started the *Zeitschrift für physikalische Chemie* (in 1887). Subsequently van't Hoff became a professor in Berlin and shifted his focus to geochemistry, where again he made pioneering contributions at the birth of a field. His stature is attested to by the fact that twenty-five years after Kolbe's attack van't Hoff was selected as the first Nobel prize laureate in chemistry. A Nobel award is an enormous honor at any time, but these days, when the Nobel Committee meets, it works on the shoulders of previous committees that have already canonized many

Figure 1. Jacobus Henricus van't Hoff (1852–1911), first Nobel laureate in chemistry. His book on the theory of the asymmetric carbon atom and optical activity was published before he obtained his Ph.D. in 1874. (Photograph courtesy of the Swedish Embassy, Washington, D. C.)

distinguished scientists. When the first Nobel Committee met, it had to pick from a galaxy of stellar candidates. It chose van't Hoff.

Kolbe was not malicious or vindictive toward van't Hoff as an individual. Rather he was the kind of individual my college laboratory mate used to call "an even-tempered fellow — always sore." For example, a few years earlier Kolbe attacked some work of the renowned French experimental thermodynamicist M. Berthelot. Of one of the latter's papers, Kolbe said that it contained "much that was new and much that was true, but the new is not true and the true is not new." He also succeeded in antagonizing some of the scientific giants of his day, such as August Kekule, attacking him as early as 1854 about priority claims for various ideas and later for his support of van't Hoff. Kekule in general maintained a dignified silence, but on occasion responded in kind, as in his comment (freely translated): "The old hen under cover of a clamorous cackle has disgorged a large turd instead of an egg."

Kolbe was not an isolated practitioner of his style of writing. In a sense he was the last member of a school that spanned the nineteenth century. An unholy trio — Berzelius, Wöhler and Liebig — was unsurpassed in the use of sarcasm and ridicule in criticism of scientific ideas or individuals they attacked. Let us pursue some of their most memorable comments.

To place the first example in proper context, I ask knowledgeable individuals to attempt to regress scientifically. Often a concept that seems eminently simple and straightforward to us today was not so viewed by our ancestors. In fact, sometimes what we believe today is the reverse of what was clearly good common sense in earlier times. As an example, I need cite only the daily motion of the sun in the sky, which brilliant intellectuals of ancient times interpreted in the inverse framework of that accepted by any schoolchild today.

I mention this general point because I was totally astonished a few years ago to learn that hardly more than a century ago yeast was not considered a living organism. In fact it was Theodore Schwann, the zoologist of the famous Schleiden and Schwann pair that formulated cell theory, who first recognized that yeast is a living organism, capable of converting sugar into alcohol. At that time, this was a disturbing suggestion received with hostility and abuse by outstanding scientists of the time — e.g., Liebig and Wöhler — who considered such an interpretation an atavistic reversion to vitalism, a view they fought vigorously. They were perfectly well aware that yeast could convert sugar to alcohol but believed yeast was inanimate and acted as other catalysts do in other chemical reactions. An unsigned article, generally attributed to Liebig, appeared in print under the title "The Secret of Alcoholic Fermentation Revealed," and it heaped ridicule upon Schwann by purporting to have made similar microscopic studies with the following results:

I have just discovered the mechanism of fermentation of wine....This discovery shows once again how simple are the means that Nature uses to produce remarkable phenomena.

If one disperses yeast in water, microscopic observation reveals little spheres, arranged in thin filaments, that are unmistakably proteinaceous in nature. If these spheres are placed in sugar water, then it becomes evident that they are animal eggs; they grow in size, undergo fission and develop into tiny organisms that multiply extremely rapidly. The shape of these animals differs from any of the 600 known types; they have the form of a distilling retort (without the cooling attachment). The spout of the stillhead is like a suction nozzle, with fine indentations internally; teeth and eyes cannot be detected; however, one can discern a stomach, an intestinal tract, an anus (a rose-red dot), a urinary organ. From the moment that they emerge from the egg, these animalcules gobble up sugar from the solution, which is very visible as it reaches the stomach. It is promptly digested and the product is unequivocally recognizable in the excrement that is subsequently evacuated...The bladder in the filled state looks like a champagne bottle, empty it appears to be a small button. With practice one can see a gas bubble grow ten-fold in size in the bladder; the bladder is emptied by a screw-like twisting of a set of muscles that form a ring around it...The excrement released in eighteen hours weighs sixty-six times as much as the animal...To summarize, sugar is sucked into the stomach of this little animal, *a stream of alcoholic fluid gushes out continuously from the anus and short bursts of carbon dioxide are squirted out by an enormous genital organ.*

If you do not believe this appeared in print, you can examine the original, in German, in the chemical journal *Annalen,* (vol. 29, p. 100) published in 1839. Liebig had no difficulty publishing this material. I should mention in passing that this journal is usually listed in library card catalogues under the entry *Justus Liebig's Annalen der Chemie.*

Some fifteen years later Liebig's scorn was transferred to Pasteur. By this time Liebig had conceded that yeast is a living organism, but he attributed fermentation to the *dead* state, to the decomposition products of animal or vegetable matter. Pasteur's view was the exact opposite, that the *living* yeast was the cause of fermentation. To confound Liebig, Pasteur introduced a trace inoculum of yeast into a medium containing only water, sugar and crystalline salts, and demonstrated that the yeast produced alcohol and simultaneously thrived and multiplied. To this demonstration Liebig countered:

> As to the opinion which explains the putrefaction of animal substances by the presence of microscopic animalcula, it may be compared to that of a child who would explain the rapidity of the Rhine's current by attributing it to the violent movement of the numerous mill wheels at Mainz.

Friedrich Wöhler (Figure 2) was also talented in this genre of criticism. His most famous contribution in this area focuses on aspects of the atomic structure of molecules. Students today think that the structural theory of chemistry, in its entirety, was revealed to Linus Pauling on Mount Pasadena, in 1936. It is not appreciated generally that modern structural concepts in chemistry are the outcome of a slow and tortuous evolution, involving many missteps, over a period of more than a century, and in fact are still being refined.

Figure 2. S. C. H. WINDLER, better known as Friedrich Wöhler (1800–1882), one of the most influential chemists of the nineteenth century. (Published with permission of the Royal Swedish Academy of Sciences.)

During the first half of the nineteenth century ideas of structure were very primitive, built upon the relatively few known and simple chemical facts. By synthesizing urea from ammonium cyanate, Wöhler showed (in 1828) that there was no distinct boundary between inorganic compounds, from mineral sources, and organic compounds, from living sources, and thereby he further confounded molecular representations. A most useful structural concept that was developed about the time was that "radicals," groups of atoms acting as a unit, maintained their integrity throughout a series of chemical transformations. These radicals, simple atoms in inorganic compounds, could be relatively complex in organic ones. For example, the "benzoyl radical," represented in modern notation as C_7H_5O, is an entity with seven carbon (C) atoms, five hydrogens (H) and one oxygen (O). It persists as such in a series of reactions that yield benzaldehyde, $C_7H_5O \cdot H$, benzoic acid, $C_7H_5O \cdot OH$, and benzoyl chloride, $C_7H_5O \cdot Cl$, respectively. In other words, the C_7H_5O unit is a very stable moiety that retains its structure despite severe chemical manipulations. Such stability became a general article of faith in structural chemistry at that time.

In the course of trying to find the cause of the acrid odor emitted by the candles during one of the royal soirées at the Tuileries palace, Jean-Baptiste Dumas discovered that in an entity such as methane (CH_4) he could progessively replace the H atoms (by treatment with chlorine (Cl) gas) and obtain, sequentially, CH_3Cl, CH_2Cl_2, $CHCl_3$ and CCl_4, respectively. He recognized that this substitution reaction had more general implications, and he and others discovered many more examples of the phenomenon. Thus a central doctrine of structural chemistry was being challenged. In his enthusiasm, Dumas overextended himself and began to interpret all types of transformations as substitution reactions even in cases for which description was absurd.

Wöhler acted as (self-appointed) God's vicar on earth obliged to strike down heresy. To make Dumas' ideas appear ridiculous, Wöhler adopted the logical trick, used often in debating arguments, of extending an opponent's position in small increments, each seemingly negligible, until you have led him over a precipice into a logical absurdity. Wöhler wrote a short article, in French, in which he outlined spurious general procedures that he had developed, on the basis of Dumas' ideas, to obtain strikingly novel compounds. This work he signed with a pseudonym.

8

He began the article by stating:

I wish to communicate promptly some very remarkable new observations in organic chemistry. These provide unforeseen, stunning verification of the theory of substitution. These experiments illustrate the exceptional import of this ingenious theory and lead me to anticipate even more striking new discoveries.

Wöhler then outlined his (bogus) experiments in the technical language of the time, which I will transpose into modern chemical terminology. He started with solid manganese acetate, which contains manganese (Mn), carbon (C), oxygen (O), and hydrogen (H) in the atomic proportions $MnC_4O_4H_6$. This material is available as pink crystals. Impressed by Dumas' principles, Wöhler exposed a sample of manganese acetate at room temperature to chlorine (Cl) gas. To his delight he obtained magnificent new crystals that were violet in color. On analysis these proved to have the composition $MnC_4O_4Cl_6$. This result, however, was not extraordinary, for the substitution of H by Cl had already been demonstrated by Dumas in other compounds. Nevertheless, encouraged by his discovery that confirmed Dumas' theory, Wöhler took the violet crystals and heated them at 110°Centigrade in the presence of chlorine gas. He was elated to find that a new solid appeared, gold in color, which on analysis proved to be $MnC_4Cl_8Cl_6$. In other words oxygen also could be replaced by chlorine. He proceeded further with the new material by dissolving it in "very pure chloral" and heating the solution to its boiling point for four days continuously in the presence of chlorine gas. He was excited to find that this treatment led to still different crystals, green silky needles. The composition of the material was $Cl_2C_4Cl_8Cl_6$, that is, manganese had been replaced by chlorine. With such a string of successes, Wöhler took a shot toward the stars. He treated the green needles even more vigorously with chlorine and observed that carbon was being liberated as carbonic acid. On subsequent cooling, a mass of crystals, leaflets in form and yellow in color, came out. Analysis showed the composition to be $Cl_2Cl_8Cl_8Cl_6$, that is, a solid form of pure chlorine had been produced from manganese acetate. As the author glowingly said:

Behold, each and every one of the elements of manganese acetate has been replaced, a perfect substitution....Credit for this discovery belongs to me....Take note in your journal.

9

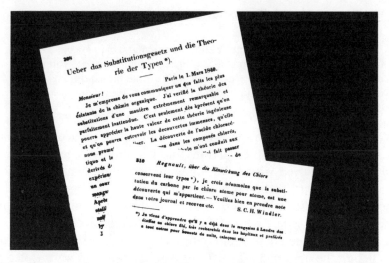

Figure 3. Excerpts from *Annalen der Chemie* showing title, signature and footnote from Wöhler's satiric article prompted by the research of Dumas.

The article was then capped by the footnote (Figure 3):

> I have just been informed that London shops now carry fabrics of spun chlorine, which are very much in demand, particularly by hospitals, in preference to other cloth for gowns, night caps, etc.

This entire literary creation was then signed by the author, S. WINDLER.

Jöns Jacob Berzelius (Figure 4) was the oldest of the unholy trio, and the father figure for the others. He was a crotchety old tryant even as a young tryant. Here is how he spoke of his countrymen:

> The main traits of the Swedish national character are laziness and dawdling...the apathy of Sweden is excessive. I must be different from my countrymen.

of a benefactor:

> A wealthy bachelor named Wilhelm Chalmers, who after he received the Wasa Order [the Swedish Order of the Garter] assumed the title of Sir William, habitually drank himself daily into such a state of intoxication that he became a bloated, pale and drooling figure. However he could not consume his entire fortune in drink, and as his end approached, Dr. Dubb persuaded him to bequeath the remains of it to public causes, among which was a fund for the establishment of a school of instruction in crafts and productive arts.

Figure 4. Jöns Jacob Berzelius (1779–1848) whose presence dominated chemistry in the early nineteenth century. He became president of the Swedish Academy of Sciences in 1810. (Published with permission of the Royal Swedish Academy of Sciences.)

and of his associates:

It is becoming increasingly unfashionable to undertake [scientific] investigations; none of my friends or former students do anything worthwhile.... Mosander has so many administrative duties and is so preoccupied with his spa that little issues from his own hands. Walmstedt diligently makes cardboard boxes himself to save a few pennies. Lynchnell has become such a drunkard that he has lost his teaching position at Uppsala...Wallquist does nothing but come in to get his salary...Professor Engstrom has always been a dolt and has now become a boozer.

He disposed of other textbook writers equally cavalierly:

Kept in bed for some days because of gout I undertook to examine textbooks other than my own.....What damned nonsense.

Contentious, acerbic old men were not confined to continental Europe. England had its share of vituperative, acrimonious scientists. An episode of special interest to Americans involves, as victim, the greatest of native American scientists, J. Willard Gibbs (Figure 5).

To chemists and physicists Gibbs is especially renowned for his elegant formulations of classical thermodynamics and of statistical mechanics. His axiomatic logical formats proved to have enormous deductive power and generality. Physicists and chemists are not often aware that during his lifetime Gibbs was far better known to mathematicians, for he was one of the major figures in the development of vector analysis and linear algebra in the form currently used universally. Contrary to a legend popular among scientists, Gibbs' accomplishments were promptly and fully recognized by his American, as well as European, contemporaries. He was elected to membership in the National Academy of Sciences in 1879 when he was only forty, an early age for election even then. (Of the four men

Figure 5. Josiah Willard Gibbs (1839–1903), greatest of native American scientists. (Published with the permission of the Department of Chemistry of Yale University and Yale University Press.)

elected that year, only the name of Gibbs is remembered by scientists today). It was particularly early for Gibbs, for at that time he had published only three papers (the last in two parts totalling 321 pages), the first appearing at the relatively late starting age of thirty-three. Obviously the exceptional import of his work was recognized almost immediately. Shortly after his election to the National Academy of Sciences, Gibbs was awarded the prestigious Rumford Medal of the American Academy of Arts and Sciences, to which he had been elected in 1880.

In the 1880s Gibbs turned his attention to problems in dynamics and from that to his formulation of vector analysis and matrix algebra. Thereby he encountered the wrath of the eminent British theoretical physicist Peter Guthrie Tait, professor of natural philosophy at Edinburgh. To Tait anything of British origin was clearly sacrosanct, superior, more fundamental or more original than an alternative concept of foreign origin. As a first-class physicist (in 1854 he had ranked first, ahead of Maxwell, in his class at Cambridge, and in 1860 he was chosen in preference to Maxwell for the physics professorship at Edinburgh) he had become personally devoted to the mathematical techniques of quaternions created by Sir William Rowan Hamilton. Hamilton (from whom also came the Hamiltonian operator of the equations of motion), was a native of Ireland, then an integral part of Great Britain.

Hamilton's genius was already evident in his precocity. At age five he knew Latin, Greek and Hebrew; at ten he was proficient in a dozen languages. He was appointed professor of astronomy, at age twenty-one, while still technically an undergraduate at Trinity College, Dublin. His achievements were so phenomenal that he was knighted at age thirty. He was the first foreigner elected to the U.S. National Academy of Sciences.

According to Hamilton, he came to the idea of quaternions, after fifteen years of cogitation, on October 16, 1843, while crossing the Brougham Bridge in Dublin. At that time, an alternative formulation of multiple algebra (called "ausdehnungslehre") had been formulated by H. G. Grassmann in Germany, and Gibbs' creations were more in the style of Grassmann than in that of Hamilton. When the methods of Gibbs in vector analysis began to displace quaternions as a mathematical tool in theoretical mechanics, Tait became defensively contentious and acerbic. Thus in the preface to his book *Elementary Treatise on Quaternions* Tait wrote: "Professor Willard Gibbs must be ranked as one of the retarders of . . . progress in virtue of his pamphlet on *Vector Analysis*; a sort of hermaphrodite monster." Subsequently in a review in *Nature* of a book by McAulay, Tait said:

It is positively exhilarating to dip into the pages of a book like this after toiling through the arid wastes presented to us as wholesome pasture in the writings of Prof. Willard Gibbs.

In another letter to the editor of *Nature,* Tait wrote:

The main object...of my present letter is to call attention to a paper by Dr. Knott, recently read before the Royal Society of Edinburgh. Dr. Knott has actually had the courage to *read* the pamphlets of Gibbs...; and after an arduous journey through these trackless jungles, has emerged a more resolute supporter of quaternions than when he entered. He has revealed the (from me at least) hitherto hidden mysteries...of Prof. Gibbs' strange symbols, Pot., Lap, Max., New, etc. The first turns out to be only the linear and vector function; and the others are merely more or less distressing symptoms of imperfect digestion.

Gibbs responded, in his characteristic dignified, cogent style, with a series of scholarly, erudite letters to *Nature,* revealing touches of subdued wit:

Indeed if my offense had been solely in the matter of notation, it would have been less accurate to describe my production as a monstrosity than to characterize its dress as uncouth.
My critic is so anxious to prove that I use quaternions that he uses arguments which would prove that quaternions were in common use before Hamilton was born.

In the mountains of gold created by Gibbs only once did a grain of dross creep in; only once did he make a mistake. That error occurred in a mathematical paper dealing with a problem of convergence of a Fourier series. The error was quickly discovered, and corrected, by Gibbs himself. Probably no scientist, except one who has published absolutely nothing, can claim such a flawless record. Tait, a prisoner of his personality, achieved a permanent niche in the history of science by including Gibbs among the persons he attacked.

Some years earlier Tait had also become involved in arguments with German scientists about who deserved primary credit for the founding of thermodynamics. Actually the answer to the question is that this set of grand principles emerged from the work and thinking of many men. In contrast to mechanics, electromagnetic field theory, or relativity, in which the names of Newton, Maxwell and Einstein, respectively, stand out preeminently, in thermodynamics, perhaps a half-dozen individuals — Carnot, Mayer, Joule, Rankine, Kelvin, Clausius — provided crucial steps for the grand synthesis. The conceptual bottle, into which early nineteenth century views of the nature and action of heat were poured, was the principle of conservation of caloric. It seems clear that the earliest of the founding six — Carnot — accepted conservation of caloric as a basic axiom in his argument (although a few individuals — Callendar, Bronsted, LaMer — claim to see an important distinction in the contexts of Carnot's use of *feu* and *chaleur*). Certainly Carnot leaned strongly on the analogy between a heat engine and a hydrodynamic one, for he said "we can reasonably compare the motive power of heat with that of a head of water."

In the water wheel, a quantity of liquid water enters at a high gravitational level and the *same* quantity of liquid water exits at the lower level, work having been produced during the fall. It was only natural, therefore, to assume that in a perfect heat engine operating in the highly abstract ideal reversible cycle created by Carnot, the heat that entered at the upper temperature level was conserved and exited in exactly the same quantity at the lower temperature, work having been produced during the thermal drop. It was from this construct that Kelvin realized one could establish an absolute temperature scale independent of the properties of any substance, and he elaborated on and expanded Carnot's analysis of the heat engine. When faced with Joule's ideas in the late 1840s, Kelvin at first rejected them (as did the Proceedings of the Royal Society when presented with one of Joule's manuscripts), because conservation of energy (work plus heat) was inconsistent with the Carnot analysis of the fall of an unchanged quantity of heat through an ideal thermal engine. In the period between 1849 and 1851, Kelvin and Clausius, each reading the other's papers closely, came to recognize that Joule and Carnot could be made concordant if one assumed that only part of the heat entering the Carnot engine at the high temperature was released at the lower level and that the difference was converted into work. Clausius was the first to express this in print. Within the next few years, Kelvin developed the mathematical expression $\Sigma(Q/T) = 0$ for "the second fundamental law of the dynamical theory of heat" and began to use the word "thermodynamic," which he had actually coined earlier. In turn Clausius's analysis led him to the mathematical formulation of $\int(dQ/T) \geqslant 0$ for the second law; and he invented the term "entropy," for, as he said: "I hold it better to borrow terms for important magnitudes from the ancient languages so that they may be adopted unchanged in all modern languages."

Tait, a contemporary, friend and collaborator with Kelvin in textbook writing, watched these developments from the sidelines, and set out to demonstrate that everything Clausius published appeared after the seminal idea or mathematical expression had already been printed by Kelvin. Thus Tait wrote,

> Clausius who published results equivalent to $[\Sigma q/t = 0]$ at a somewhat later date calls such a term as q/t the equivalence-value...and does not...even refer to the absolute definition of temperature.

In his later years (*Thermodynamics,* 1877) he may have acquired some self-insight for he wrote:

> It is possible that circumstances may have led me to regard the question from a somewhat too British point of view....But even supposing the worst, it appears to me that unless contemporary history be written with some little partiality, it will be impossible for the future historian to compile from the works of the present day a complete and unbiassed statement. Are not both judge and jury greatly assisted to a correct verdict by the avowedly partial statements of rival pleaders? If not what is the use of counsel?

Special pleading with chauvinist overtones has appeared in many large countries of modern Europe. The Frenchman P. Duhem (Figure 6), whose career partly overlapped Tait's, was even more cosmic and outrageous in his claims and criticisms. In his famous philosophical monograph on the foundations of theoretical physics, Duhem devoted substantial space to the logical structure of the creations of great French theoreticians such as Descartes, Laplace, Poisson, Ampere, Fourier, and waxed eloquent on their constructions:

> such an edifice [mechanics] stood as the perpetual ideal of abstract intellects, *especially of the French genius.* In pursuing this ideal it has raised monuments whose simple lines and grand proportions are still an object of delight and admiration....
> Not at all like this is the case of *the ample but weak mind of the English physicist.* Theory is for him neither an explanation nor a rational classification of physical laws, but a model of these laws, a model not built for the satisfying of reason but for the pleasure of the imagination. Hence it escapes the domination of logic.... To a physicist of the school of Thomson [Kelvin] or Maxwell, there is no contradiction in the fact that the same law can be represented by two different models. Moreover, the complication thus introduced into science does not shock the Englishman at all; for him it adds the charm

Figure 6. Pierre-Maurice-Marie Duhem (1861–1916), theoretical physicist, historian and philosopher of science. Professor at Bordeaux for most of his professional life (1894–1916). Behind this benign expression lay a contentious, acrimonious personality with an uncommon talent for creating professional enemies and for misjudging the potential of the revolutionary concepts that appeared at the dawn of the twentieth century (Maxwellian electromagnetism, atomistic and quantum physics, relativity). To his deep chagrin he was never called to a professorship in Paris, although he was elected to the Academy of Sciences. (Photograph reproduced through the courtesy of the French Academy of Sciences.)

of variety. His imagination...does not know our need for order and simplicity.... Thus in English theories we find those ⊾isparities, those incoherencies, those contradictions, which we are driven to judge severely because we seek a rational system where the author has sought to give us only a work of imagination.... The English type of theory does not subject itself...to the rules of order and unity demanded by logic....The English mind is characterized by the meagre way it makes use of abstractions and generalizations....The reasoning and calculation by which Maxwell tried on several occasions to justify [his electrodynamic equations] abound in contradictions, obscurities, and plain mistakes....Maxwell's *Treatise on Electricity and Magnetism* was in vain attired in mathematical form. It is no more of a logical system than Thomson's [Kelvin's] *Lectures on Molecular Dynamics*...[Each] consists of a succession of models....*A gallery of paintings is not a chain of syllogisms.*

17

There is some validity in such a viewpoint toward the atomistic mechanical dynamical theories of Lord Kelvin in his later years, when in fact he expressed his philosophical views in the well-known aphorism: "It seems to me that the test of 'Do we or do we not understand a particular point in physics' is 'Can we make a mechanical theory of it.'"

However, Duhem's criticism is hardly appropriate with respect to classical thermodynamics, in which an expansive logical structure was constructed by Kelvin on the foundation of two abstract, non-mechanical axioms: the two-fundamental laws of energy and entropy. The criticism was also totally inappropriate with regard to the electromagnetic field theory of Maxwell, which Duhem continually criticized pitilessly, probably because he was a partisan of Helmholtz's alternative formulation.

Duhem also was confronted with the name of the premier English theoretical scientist of all time — Newton. This giant he grudgingly admired:

> Surely Newton yielded nothing either to Descartes or to any of the great classical thinkers in the ability to offer very abstract ideas with perfect clarity and very general principles with great accuracy.

Nevertheless, he did not hesitate to cite disparaging comments from some of Newton's contemporaries:

> The Newtonians who endow material elements with attractions and repulsions acting at a distance seem...to be adopting one of those purely verbal explanations usual with the old Scholasticism. Newton's *Principia* had hardly been published when his work excited the sarcasm of...Huygens. "So far as concerns the cause of the tides given by Mr. Newton," Huygens wrote Leibniz, "I am far from satisfied, nor do I feel happy about any of his other theories built on *his principle of attraction, which to me appears absurd.*"
> If Descartes had been alive at the time, he would have used a language similar to that of Huygens. In fact, Father Mersenne had submitted to Descartes a work by Roberval in which the author adopted a form of universal gravitation long before Newton.... Descartes expressed his opinion as follows: *"Nothing is more absurd than the assumption---that a certain property is inherent in each of the parts of the world's matter and that, by the force of this property, the parts are carried toward one another and attract each other."*

It is ironic that when Duhem was touting French science at the close of the nineteenth century, its Periclean period had passed and the dominant position of French theoreticians of a century earlier had been lost, first to British and then to German scientists. This was particularly obvious at the turn into the twentieth century when Poincaré and Langevin were overshadowed by Boltzmann, Planck and Einstein. Such a decline, coming on top of the humiliating peace treaty of 1871 imposed by the Prussians on the French, may have exacerbated Duhem's hurt pride, as it did that of other French scientists such as Pasteur, and caused him to express himself even more stridently. He always had a reputation as a contentious and acrimonious man. Like Tait and Kolbe, Duhem was politically conservative, generally five miles to the right of Genghis Khan. Starting with a rigid, ultraconservative Catholic royalist view and taking strong anti-Republican, anti-Semitic, anti-Dreyfus, anti-Maxwell, and anti-relativity stances, he succeeded in antagonizing even many of his influential French contemporaries. The writing of books such as, *Chemistry, Is it a French Science?*, did not endear him to non-French scientists. Nor did comments such as the following:

> At no time have French or German physicists by themselves...reduced physical theory to being nothing but a collection of models...; it is an English importation...[a] taste for the exotic, the desire to imitate the foreign, the need to dress the mind as well as the body in the fashion of London.... Moreover, the loud admiration for the English method is for too many a means of forgetting how little apt they are in the French method, that is, how difficult it is for them to conceive an abstract idea and to follow a rigorous line of reasoning. Deprived of strength of mind, they try, by taking on the outward ways of the ample mind, to make one believe they possess intellectual amplitude...it is easy to ape the defects of foreigners.

The English were a favorite target of other continentals also. Among Berzelius's writings one finds the following:

> The chemists of England live in their own world oblivious to what is being done in France and totally unfamiliar with work in Germany....The commercial instincts of the English nation find expression even among their intellectuals. There is a great deal of litigation about priority in the most trivial matters, and despite the fact that they get along fairly well with each other, they seem like puppies who stand and snarl over bones from which the meat has been gnawed on the continent.

As Liebig once said, but apropros of others,

> It is horrible to think that the time must come when we believe only in the past as true and regard the future as froth and fantasy...because we can no longer keep up.

Even "Young Turks," with the passage of time, turn into "Old Turks."

Anonymous (attributed to J. Liebig) (1839). "The Secret of Alcoholic Fermentation Revealed", *Annalen der Chemie,* Vol. 29, pp. 100-104.

Brougham, H. P. Reviews of "The Bakerian Lecture on the Theory of Light and Colours" by Thomas Young. *Edinburgh Review* (1803), Vol. 1, pp. 450-460; (1804) Vol. 5, pp. 97-103.

Cardwell, D. S. L. (1971). *From Watt to Clausius,* Cornell University Press, Ithaca, New York.

Carnot, S. (1824). *Reflexions sur la puissance motrice du feu,* Bachelier, Paris.

Duhem, P. (1906). *The Aim and Structure of Physical Theory,* Chevalier and Riviere, Paris. Translated by P. P. Wiener, 1954, Princeton University Press, Princeton, N. J.

Gibbs, J. W. (1893). "Quaternions and Vector Analysis", *Nature,* Vol. 48, pp. 364-367; see also (1891) issue, vol. 43, pp. 511-513, (1891) issue, Vol. 44, pp. 79-82.

Gillispie, C. C. (1960). *The Edge of Objectivity,* Princeton University Press, Princeton, N. J.

Harden, A. (1911). *Alcoholic Fermentation,* Longmans, Green and Co., London.

Hoffmann, B. and Dukas, H. (1972). *Albert Einstein, Creator and Rebel,* Viking Press, New York.

Jorpes, J. E. (1966). *Jac. Berzelius, His Life and Work,* Almqvist and Wiksell, Stockholm.

Kekule, A. (1965). *Cassirte Kapitel aus der Abhandlung: Über die Carboxytartronsäure und die Constitution des Benzols,* Verlag Chemie, GmbH, Weinheim/Bergstr., Germany. Published on the occasion of the Centennial Jubilee of the ring structure of benzene.

Kolbe, H. (1873). "Salts of Alkylsulfuric Acids". *Journal für praktische Chemie,* Volume 115, pp. 266-268.

Kolbe, H. (1877). "Signs of the Times". *Journal für praktische Chemie,* Volume 15, pp. 473-477.

Moon, P. and Spencer, D. E. (1965). *Vectors,* Van Nostrand, Princeton, N. J.

Tait, P. G. (1877). *Sketch of Thermodynamics,* second edition (first edition 1868), David Douglas, Edinburgh.

Tait, P. G. (1893). "Quaternions as an Instrument in Physical Research". *Nature,* Vol. 47, pp. 225-226; Vol. 49, pp. 193-194.

van't Hoff, J. H. (1874). *Vorstel tot uibreiding der tegenwoordig in de scheikunde gebruikte structure-formules in de ruimte,* Greven, Utrecht.

van't Hoff, J. H. (1875). *La Chimie dans l'Espace,* Bazendijk, Rotterdam.

Wheeler, L. P. (1951). *Josiah Willard Gibbs,* Yale University Press, New Haven, Conn.

Windler, S. C. H. (pseudonym for F. Wöhler) (1840). "The Law of Substitutions and the Theory of Types". *Annalen der Chemie,* Vol. 33, pp. 308-310. See representation in Figure 3.

II
The Clouded Crystal Ball:
Creases of the Mind

Let us turn now to another aspect of distorted perception, one that does not reflect so strongly the frustrations of old men, but that highlights examples of clouded judgment based on sound experimental or theoretical grounds, that is, sound in terms of knowledge and understanding in their times.

Most people have heard of proofs, purportedly produced by physicists, that an airplane in principle would never be constructed, that is, actually made to fly. Some supposedly appeared shortly before the Wright brothers launched their first successful flight. Whenever I heard any of these stories, I assumed they were merely calumnies invented and promulgated by engineers to embarrass physicists of whom they are deeply envious. But I have tracked down some written pronouncements on this subject. It seems likely that even Newton believed a self-propelled plane impossible, and certainly Lord Kelvin (Figure 1) was convinced no airplane would ever fly. Here is Kelvin's comment: "I have not the smallest molecule of faith in aerial navigation other than ballooning, or of expectation of good results from any of the trials we hear of."

This assessment was based on sound principles of fluid mechanics as understood at that time. In particular, oblique pressures on a wing were evaluated from a Newtonian equation that incorrectly gives only 5–10 percent of the true values at acute angles of incidence. The details of the mistake are not essential for our purpose. What is pertinent is that the man regarded in his lifetime as probably the greatest physicist of the nineteenth century applied sound principles

Figure 1. Lord Kelvin (William Thomson, 1824–1907) in his sixties. (From *Life of Lord Kelvin* by S. P. Thompson.)

and logic of physics in his analysis and came to an apparently unassailable conclusion, which in the course of time proved to be wrong.

In contrast to Kolbe or Berzelius, Kelvin (born William Thomson) was truly an even-tempered person. Furthermore, he was an immensely gifted and talented individual. He enrolled in the University of Glasgow at age ten; he published his first paper, in the Cambridge Mathematical Journal, at the age of seventeen, at which time he also entered Cambridge University. He was appointed professor of natural philosophy (as physics was then called) at Glasgow, in 1846, when he was twenty-two years old. Before he was thirty, he (in parallel with Clausius in Zurich) had built the foundations of classical thermodynamics and established a framework for its manifold ramifications. He also made immense contributions to many other areas of theoretical physics, including electromagnetism, mechanics and hydrodynamics. He was obviously a smart professor — he was a rich professor. He was versatile and ingenious in many areas of engineering, particularly electrical, and because of his skill in practical matters earned many thousands of British pounds annually as a consultant, at a time when a pound was equivalent to five dollars and had a purchasing power many, many times greater than today. (For example, a man of the stature of Gibbs was paid two thousand dollars a year by Yale, and that only after Johns Hopkins University almost stole him away.) And Kelvin knew how to spend money. He spent large parts of each year on his 126 ton yacht, which required a crew of over a dozen, and his home in Glasgow was probably the first in the world to have all-electric lighting. He also built a mansion (Figure 2), Netherhall, in Largs (a seaside town thirty miles west of Glasgow) suitable for a lord of the realm. He was elected a fellow of the Royal Society at age twenty-seven, and knighted at age forty-two, the latter honor being bestowed in recognition of his senior scientific-engineering role on the famous ship, the Great Eastern, during the laying of the first successful trans-

Figure 2. Netherhall, Lord Kelvin's seaside home (near Largs), which he inhabited in addition to his home in Glasgow, where he was professor of natural philosophy throughout his academic lifetime. (From *Life of Lord Kelvin,* by S. P. Thompson.)

atlantic telegraph cable. In his more mature days he received many scientific, academic and public honors. He was elected to membership (and in some cases to high office) in almost a hundred societies, including the National Academy of Sciences of the United States, the Academia Svecica (Sweden), the Hungarian Academy of Sciences, the Accademia dei Lincei (Rome), the Institut de France, the Imperial Academy of Sciences (St. Petersburg), etc. He was awarded many scientific medals, foreign and domestic, including the Copley Medal of the Royal Society (in 1893); he received more than twenty honorary degrees including an LL.D. from Yale University (in 1902, evidently on the recommendation of Willard Gibbs), and he was elevated to the peerage in 1892. On the last occasion, William Thomson adopted the name "Kelvin" to link his name with the Kelvin River, which flowed in the grove below the buildings of the University of Glasgow, with which he was associated throughout his entire adult life. His stature in the eyes of the British people is attested to by his culminating honor, burial in Westminster Abbey — right next to Newton.

Kelvin was a prolific writer and a man with an enormous range of interests. He published two dozen books, over 600 papers and almost uncountable book reviews, letters and speeches. In addition he was the inventor in some seventy patents. He mined mounds of pure gold, but inevitably, a few nuggets of rust. Picayune critics, from the vantage point of a century later, have been able to separate out some of the slag.

Because he spoke out so often on the subject, Kelvin's most notorious conservative stance was that against evolution and the doctrine of uniformitarianism in geology, with which it was closely coupled. Although he attended churches of both persuasions, he was not a rigid Presbyterian or Episcopalian. In fact, his view that life on earth originated from spores coming in from outer space, was anathema to the dominant clergy. He felt that "it was impossible that atoms of dead matter should come together so as to make life." His personality is revealed by stands such as his opposition to the granting of degrees by Cambridge to women, his refusal to accept the newly-coined word "physicist" (he called it a "vulgarism"), although he found "physics" perfectly acceptable, his consistent avoidance of the term "entropy" throughout his thermodynamic writings (he used the term "dissipation of energy").

From an early age Kelvin was interested in geology and methods of estimating the age of the earth. On the occasion of his induction to the professorship at Glasgow in 1846 he presented a dissertation on the "Age of the Earth and its Limitation as determined from the Distribution and Movement of Heat within it" in which, he remarked later, he had "proved the untenability of the enormous claims for time which... geologists and biologists had begun to make and to regard as unchallenged."

His analyses of the age of the earth were expanded further, particularly in the early 1860s, immediately after the appearance of Darwin's *Origin of Species*. Kelvin considered Darwin's views on the absence of Design in Nature unscientific. In 1865 he wrote a paper with the pointed title "The Doctrine of Uniformity in Geology Briefly Refuted." When Kelvin the Scot said "briefly," that is precisely what he meant. Here is the entire text; it consists of four sentences.

> The "Doctrine of Uniformity" in Geology, as held by many of the most eminent of British geologists, assumes that the earth's surface and upper crust have been nearly as they are at present in temperature, and other physical qualities, during millions and millions of years. But the heat which we know, by observation, to be now conducted out of the earth yearly is so great, that if *this* action had been going on with any approach to uniformity for 20,000 million years, the amount of heat lost out of the earth would have been about as much as would heat, by 100° Cent., a quantity of ordinary surface rock of 100 times the earth's bulk. (See calculation appended.) This would be more than enough to melt a mass of surface rock equal in bulk to the *whole earth*. No hypothesis as to chemical action, internal fluidity, effects of pressure at great depth, or possible character of substances in the interior of the earth, possessing the smallest vestige of probability, can justify the supposition that the earth's upper crust has remained nearly as it is, while from the whole, or from any part of the earth, so great a quantity of heat has been lost.

Also interesting are Kelvin's arguments dealing with the age of the sun. He started with the first law of thermodynamics: if the sun is continuously losing energy by radiation, then (a) its temperature must be and must have been dropping or (b) new energy must be supplied by the capture of meteors or by gravitational contraction. His calculations then led him to conclude that the sun must have been substantially hotter a million years ago and that greater variations in the earth's climate must have taken place in antiquity than the uniformitarian school had assumed. It is also possible to calculate the age of the sun if its energy comes from gravitational contraction. Kelvin used several formulations of this problem including the one worked out by his friend Helmholtz, which proceeds as follows.

When matter at the surface of the sun falls toward the center, it gains potential energy of 1.91×10^{15} ergs per gram. (To obtain this figure, it was necessary to know the mass of the sun (1.99×10^{33} grams), its radius (6.96×10^{10} centimeters) and the gravitational constant (6.67×10^{-8}), all of which were available). According to a fundamental theorem of mechanics, half of this energy could be converted to radiation (the other half becoming kinetic energy). Current measurements show that in each second the energy radiated by the sun is 1.96 ergs per gram. Therefore, if the energy radiated by the sun arises from gravitational contraction, there is enough energy for $1.91 \times 10^{15}/1.96$ or approximately 10^{15} seconds, that is thirty million years. This is much, much shorter than the age of billions of years assumed by geologists and evolutionists.

The logic and premises of this argument were essentially unassailable in the 1860s. Radioactivity was not discovered until almost forty years later. That it was associated with enormous energy changes was not appreciated for another decade. That the sun's energy could be accounted for by nuclear fusion was not really established until Hans Bethe did so around 1930. Thus in 1860 there was no reasonable alternative to gravitational contraction as the source of the sun's energy. So Kelvin's and Helmholtz's premise was unassailable and their calculations impeccable.

Nevertheless, some biologists and geologists argued back. Certainly the most effective of these was T. H. Huxley, Darwin's scientific lawyer in all the controversies raging at that time. An immensely articulate controversialist and a master of dialectical argument, Huxley examined the different lines of mathematical assessment that seemed to limit the age of the earth to about a hundred million years and then formulated one of his famous passages:

> But I desire to point out that this seems to be one of the many cases in which the admitted accuracy of mathematical processes is allowed to throw a wholly inadmissible appearance of authority over the results obtained by them. Mathematics may be compared to a mill of exquisite workmanship, which grinds you stuff of any degree of fineness; but, nevertheless, *what you get out depends on what you put in; and as the grandest mill in the world will not extract wheat-flour from peascods, so pages of formulae will not get a definite result out of loose data.*

Kelvin was tenacious, but invariably courteous and patient. Whereas many physicists would have said "Wouldn't it be nice if biologists learned some physics," Kelvin's response was:

> Though a clever counsel may, by force of mother-wit and common sense, aided by his very peculiar intellectual training, readily carry a jury with him

to either side when a scientific question is before the Court, or may even succeed in perplexing the mind of a judge, I do not think that the high court of educated scientific opinion will ever be satisfied by pleadings conducted on such precedents.

Huxley could be very stinging in debate, as in his famous interchange with the bishop of Oxford, Samuel Wilberforce ("Soapy Sam"), his most famous public antagonist in arguments about evolution and the descent of man, and a formidable, witty and charismatic disputationist. As remembered by listeners present at the famous 1860 meeting of the British Association for the Advancement of Science in Oxford, the following exchange took place.

Wilberforce (turning to Huxley, with sardonic courtesy): I beg to know, was it through your grandfather or your grandmother that you claim your descent from a monkey? *Huxley (after explaining Darwin's ideas and then exposing the Bishop's ignorance and vacuity):* I would not be ashamed to have a monkey for my ancestor, but I would be ashamed to be connected with a man who used great gifts to obscure the truth.

However, in contrast to his attitude toward Bishop Wilberforce, Huxley held Kelvin in the highest esteem, scientifically and personally. In the very midst of their controversy, Huxley had to introduce Kelvin as his successor to the chair of President of the British Association for the Advancement of Science, and he used these words of tribute and warmth:

Permit me, finally, to congratulate the Association that its deliberations will be presided over by a gentleman, who in spite of what I must call the trifling and impertinent accident of birth, is to all intents and purposes a Scotchman. For it is in Glasgow that Sir William Thomson [Kelvin] has carried out, now for five-and-twenty years, that series of remarkable researches which...places him at the head of those who apply mathematics to physical science, and in the front rank of physical philosophers themselves — an achievement, which, in this age of the cultivation of science, and in the pressing rivalry of able and accomplished men in all directions, confers upon him who realizes it the title of an intellectual giant. On the one hand Sir William Thomson has followed out to the utmost limit of...scientific imagination those speculations which carry men further forward in the pursuit of truth; and, upon the other hand, he has, by the versatility which is conferred only on the ablest men, been able to turn his vast knowledge and his remarkable ingenuity to the perfection and the carrying out of...those great feats of engineering which may be looked upon as among the great practical triumphs of this age and generation. Those are the public, notorious, and obvious feats of your President-elect. What is less known and less obvious — his personal qualities — are such as I dare not, and will not, here dwell upon; but upon one matter which lies within my own personal knowledge I may be permitted to say of him, as the old poet says of Lancelot, that

> "Gentler knight
> There never broke a lance."

Seeing the future in a clouded crystal ball is not limited to chemists and physicists. One of the most celebrated examples in science comes from biology, and revolves around Gregor Mendel (Figure 3), the obscure monk who discovered the fundamental laws of heredity. In part the complete neglect of this work for thirty-five years came from sheer oversight (the results were published in an obscure journal in the hinterlands of Europe), but in part it also was due to the failure of what is now fashionably called "The Scientific Establishment" to appreciate the significance of Mendel's discovery. The oversight is usually described in biology texts, or in widely-read scientific journals such as Time or the Wall Street Journal, so I shall touch on it only in passing. However, I would like to elaborate on the reception Mendel received from the people who learned of his

Figure 3. Johann Gregor Mendel (1822–1884), substitute teacher of physics and natural history at the Brunn Technical School. In his lifetime he was more widely known for his meteorological studies than for his hybridization experiments. (Photograph from *Gregor Johann Mendel, Leben, Werk and Wirkung* by H. Iltis, Berlin, 1924. Translated by E. and C. Paul, W. W. Norton and Co., New York, 1932.)

work. Relevant to this aspect is a little of Mendel's personal history; so let me recite a few biographical details.

Mendel was born in 1822 in the small town of Heinzendorf, in Moravia, then a part of the Austrian empire, although now in Czechoslovakia. The village had a mixed population of Germans and Czechs. He came from peasant stock, and was the only son in a large poor family. In those days in central Europe the church recruited its priests largely from the poorer economic classes. Even in this century, a Dietrich Bonhoeffer, one of the few German martyrs of the Nazi era, was one of the rare members of the clergy who came from a wealthy family with high status in society. Mendel may very well have been encouraged by his father, as he certainly was by his physics teacher, to enter the priesthood so that the limited means of the family would have one less human demand on it. In any event, late in 1843 young Mendel entered the Augustinian monastery in Altbrunn, at which time, as a novice, he adopted the name Gregor to replace his christened Johann. After four years of clerical training, he was ordained a priest, in 1848. At that time, the Augustinian order staffed elementary schools in the Austrian empire, and Mendel was assigned a position as a substitute teacher in a high school. To become a regular teacher, however, he had to take a state examination for certification. He took such an examination in 1850, and failed.

About fifty years later, another young German, A. Einstein, took a similar examination, and he too failed. From these incidents, I have concluded that the German examination system is a remarkably effective device for detecting geniuses.

Having failed his examination, Mendel was sent off to graduate school, to the University of Vienna. There he studied physics (with the famous Doppler, as well as the well-known theoretical physicist, Ettinghausen), mathematics, biology and paleontology. He acquired mathematical and quantitative experimental skills that must have been extremely unusual for a botanist in the middle of the nineteenth century and that surely colored the design and interpretation of his subsequent experiments. He returned to Brunn in 1853, and shortly thereafter filled an assignment as a teacher (of physics and natural history), a position that he held until 1868.

By the spring of 1856 he evidently found enough time to start the now famous series of experiments with peas. His objective was later very clearly stated: to follow the progeny of hybrids

in such a way as to make it possible to determine the number of different forms under which the offspring of hybrids appear; to arrange these forms

29

with certainty according to their separate generations; definitely to ascertain their statistical relations.

Such questions had never been asked before; hence an experimental program had to be designed to discover any statistical rules.

Mendel's experiments over the following eight years (Figure 4) led him to the formulation of the laws of segregation and combination, and of independent assortment.

These principles, and the experiments upon which they were based, were described by Mendel in two lectures that he gave in the spring of 1865 to the Brunn Society of Natural History. These lectures were combined into a single paper that was published in 1866 in the *Proceedings* of this small backwater group, under the unassuming title of "Experiments on Plant Hybridization." The contents, using combinatorial statistics and mathematics that now seem simple and can be recognized as a rudimentary forerunner of matrix algebra, surely did not appeal to botanists of the middle of the nineteenth century, either in the spoken or written form. Nevertheless, had anyone been interested, he or she could have seen the published article

Figure 4. The garden where Mendel carried out his research on hybridization of peas by artificial pollination. From 1856 to 1863 he cultivated some 30,000 plants and kept meticulous records of their characteristics. (Photograph from *Gregor Johann Mendel, Leben, Werk and Wirkung* by H. Iltis, Julius Springer Verlag, Berlin, 1924.)

in the archives of any one of the 120 learned societies, academies and libraries to whom the Brunn Society of Natural History sent its *Proceedings*. Ultimately this issue (Volume 4) of the *Proceedings* was discovered in twenty-two American libraries, and there were at least two copies in the city of London. Nevertheless, before 1900, there were only five printed references to Mendel's paper, two being essentially encyclopedic, and none showed any appreciation of the significance of the work.

Scientists, like others, are vain people. When his paper is published, a scientist orders reprints of it, and then sits around waiting for fan mail — reprint requests. Knowing that everyone will be anxious to read my work as quickly as possible, I usually order a thousand reprints, and am chagrined when the expected flood of fan mail turns out to be a trickle of one or two postcards. Mendel was a much more modest person. According to the records of the editor of the *Proceedings*, Mendel ordered only forty reprints. He received *no* reprint requests. Nevertheless, he sent out two, unrequested: one to Professor Carl von Nageli, the professor of botany at Munich (the Harvard of the German-speaking world) and another to Professor Anton Kerner von Marilaun, at Innsbruck, the outstanding authority on plant hybridization in the Austrian empire.

After Kerner's death, his assistant (Ginzberger) found Mendel's reprint in the professor's library. The edges of the pages had never been cut; clearly the reprint had never been opened. It resides now in the library of the University of Uppsala.

The copy to von Nageli was transmitted with a very deferential letter from the descendant of peasants to the aristocratic professor who was the outstanding authority on hybridization. It is clear that von Nageli failed totally to appreciate Mendel's discovery, and certainly never referred to this work in his own publications. He was courteous enough, however, to respond to Mendel, making some essentially supercilious criticisms. Furthermore, he encouraged Mendel to make a detailed study of *Hieracium* (hawkweed). For reasons that could not have been guessed at the time, if one wanted to choose the *world's most unsuitable material for the study of inheritance,* one should choose hawkweed. It was not realized then that the seeds of hawkweed are purely maternal in origin, that is do *not* originate from meiosis of cells followed by fertilization, but instead are generated by a process called apomixis, somewhat like parthenogenesis. Poor, modest Mendel followed the advice of the renowned professor and stumbled on into a disappointing and blind alley.

At about this time (1868) Mendel was elected abbot of the Brunn monastery. That was somewhat akin to becoming a dean in a university; his productive scientific work stopped quickly. He did publish three more papers, only one of which (in 1870) was in genetics, but his heavy administrative and public duties occupied most of his time. He died, of Bright's disease, in 1884, totally convinced of the general validity of the principles he had formulated, despite the unexplainable contradiction in the work with *Hieracium*. In this respect he strongly resembled Einstein who, a generation later, was similarly certain that the principles of special relativity were correct despite seemingly contradictory experiments (not resolved for a decade) by a highly respected experimentalist, Walther Kaufmann, on the electrodynamic properties of the electron. However, unlike Einstein's work, which was recognized and appreciated almost immediately after publication by the great theoretical physicists of his time (e.g. Planck, Lorentz), Mendel's contribution remained essentially buried until it was rediscovered in 1900 by De Vries, by Correns (a student of Nageli) and by Tschermak.

It is ironic that the subsequent development of genetics continued to be plagued by repeated attacks and misperceptions based on different philosophical or ideological grounds. During the period of Mendel's oblivion, remarkable discoveries had been made in the cytology of fertilization and cell division and in the role played by chromosomes. These, together with increasing awareness of discontinuous variations in inheritance and the rise of the concept of germ cells, as distinguished from somatic cells, provided a very fertile environment for the reception and wide dispersal of Mendel's ideas. The most active proponent of these ideas at the very outset of the twentieth century was William Bateson, at Cambridge, who in fact introduced terms such as *genetics* and *zygote,* and contributed very extensively to the experimental and conceptual flowering of the field. Nevertheless, when it became overwhelmingly evident, particularly from the classical studies of T. H. Morgan's group at Columbia University, that parallel relations being found between genes and chromosomes were not just accidental, Bateson took a very antagonistic stand against the chromosome mechanism. In 1916 he wrote:

> It is inconceivable that particles of chromatin or of any other substance, however complex, can possess those powers that must be assigned to our factors [genes]....The supposition that particles of chromatin, indistinguish-

able from each other and indeed almost homogeneous under any known test, can by their material nature confer all the properties of life surpasses the range of even the most convinced materialism.

It is of interest to contrast this passage with one written twenty years earlier (in 1895), long before the evidence was in, by another famous biologist, E. B. Wilson, who prophetically remarked:

Now chromation is known to be closely similar to, if not identical with, a substance known as nuclein...which analysis shows to be a tolerably definite chemical compostion of nucleic acid (a complex organic acid rich in phosphorous) and albumin. And thus we reach the remarkable conclusion that inheritance may, perhaps, be effected by the physical transmission of a particular chemical compound from parent to offspring.

Shortly thereafter Wilson even suggested that the nucleic acid is the component transmitted hereditarily.

Attacks on Morgan continued for many years after 1916. A decade later E. C. Jeffrey of Harvard criticized the cytological studies of the fruit fly and stated that

fundamental reservations must be made, in any general conclusions, regarding the origin of species, and the laws of heredity, which can be drawn from the experimental study of this species [*Drosophila*].

These erroneous results, which have apparently been reached as a consequence of too superficial study in the case of *Drosophila melanogaster,* present an excellent example of the dangers of what may be called the *in vitro* or purely experimental study in the biological sciences. A number of biologists are apparently of the opinion that studies carried on in glassware, greenhouses, or laboratories have a fundamental value and transcend in importance other kinds of biological investigation. It is necessary, however, to correlate the results seen at best, but darkly, in the glass experimental houses with those obtained by the study of living matter in the open, in other words, in nature. This was essentially the method pursued by Charles Darwin, who compared the only experimental evidence in general available in his time, namely, that furnished by cultivated plants and domesticated animals, with the conditions presented by plants and animals in nature. A large part of the perennial value of the "Origin of Species" is the result of this broad and solid method.

The extreme experimentalists, moreover, have apparently forgotten a very old and extremely prudent adage, namely, that those who live in glass houses should not throw stones. Professor Morgan in his "Critique of the Theory of Evolution," which has enjoyed a tremendous vogue in recent years, *has damned with faint praise or attempted to controvert many of the fundamental principles of the biological sciences and has attempted to set up in place of them an evolutionary hypothesis based on the study of a single aberrant species of Drosophila.* There is an interesting contrast between the Morgan hypothesis and that of Charles Darwin. The latter was able to buttress his views with the conclusions reached by morphologists, paleontologists, embryologists and biogeographers. In spite of the strong support supplied by the general biological situation in Darwin's time his hypothesis met with the bitterest opposition. The Morgan hypothesis of mutation based on the study of *Drosophila melanogaster* by contrast runs counter to practically all the inductive conclusions of the biological sciences. In contrast to the Darwinian hypothesis, moreover, it has been acclaimed at once by almost the entire body of biologists. *The history of science appears to warrant no expectation of long life for the mutation hypothesis.* It is, moreover, inconceivable that a science which has reached as its supreme achievement the theory of evolution should itself progress by unreasoning revolution and the subversion of the fundamentals of the biological sciences. It is in fact not impossible that *before many years have elapsed the doctrine of mutation will appear to the eyes of men as a fantastic Fata Morgana, appropriately staged on the exaggerated skyline of the lower Hudson* [i.e., at Columbia University].

The argument used by Jeffrey is a type that recurs frequently in biology (and elsewhere): one must use a holistic approach, not a reductionist one, in examining a fundamental problem of broad significance. In the same period and general context it was expressed even more succinctly by the famous biologist J. G. Needham:

> Dr. Howard suggests that we give more time to taxonomy and ecology and less to physiology and genetics. This is a good suggestion. We are all out of balance. Some of our laboratories resemble up-to-date shops for quantity production of fabricated genetic hypotheses. Some of our publications make a prodigious effort to translate everything biological into terms of physiology and mechanism — an effort as labored as it is unnecessary and unprofitable.

More recently this type of argument was used by some classical physiologists to castigate biochemists and molecular biologists for tearing apart cells to isolate and study the clearly defined molecular species isolated therefrom. The conservatives argued that the behavior of a pure enzyme or nucleic acid in a simple solution in a test tube has no relation to its role in the complex environment of a living cell. Logically they are correct — there need be no relation, and one must be critical and alert. Experience has shown, however, that the reductionist approach, slowly and carefully integrated, has discovered principles which do in fact operate in the intact cell. Likewise the proper approach to the study of genetics in man has turned out to be to study peas.

Anti-Mendelism-Morganism reached its acme in the Soviet Union during the three decades from 1930 to 1960. This story is a long and sordid one. For our present purpose, it will suffice to record a few of the remarks of the adherents to the Marxist genetics of the "great" Soviet biologist, T. D. Lysenko. The heart of his position was that there are two diametrically opposed biologies: (a) the progressive, communist, materialistic Michurinist-Lysenkoist one; (b) the bourgeois, capitalist, reactionary, idealistic, Weismannist-Mendelist-Morganist position. Here are some of the perceptive remarks of a few of the expositors of Soviet genetics, in the order of their increasingly incisive intellectual insight.

Our native geneticists, those attempting to defend the "truths" of Mendelism-Morganism, should take pause over the significant fact that the philosophical foundations of the theory they defend had already found a place in the history of pseudoscience...exposed by Engels....The struggle against the remnants of bourgeois opinions in science, the implacable struggle against pseudoscience and the idealistic and metaphysical distortions, is the business of every scientist and every scientific institution of our land.

The long-term struggle between the two trends in biology has irrefutably demonstrated that the *Mendel-Morgan trend in biology is a reactionary, antinational trend, and that it is impeding the further development of biological science* and is causing great harm in practice to socialist agriculture. Bourgeois genetics has become the fashionable "science" abroad, propagandizing "eugenics" and race politics. *Weismannism-Morganism serves today in the arsenal of contemporary imperialism as a means for providing a "scientific base" for its reactionary politics.*

The Weismannist-Mendelist-Morganist current in biology is an antinational, pseudoscientific, deleterious current. It disarms practice and orients man toward resignation to the allegedly eternal laws of nature, toward passivity, toward an aimless search for hidden treasure and expectation of lucky accidents. The bourgeoisie is interested in promoting Weismannism, which assumes a political significance through eugenics and various race "theories." Weismannist (Mendelist-Morganist) genetics is a spawn of bourgeois society, which finds the recognition of the theory of development unprofitable because, from it, in connection with social phenomena, stems the inevitability of collapse of the bourgeoisie. Bourgeois society prefers the "theory" of immutability of the old, of appearance of something new only from recombination of the old or by happy chance. This "theory" leads to a passive contemplation of supposedly eternal phenomena of nature, to a passive expectation of accidental variation. That is why Mendelist-Morganist genetics is held in such great esteem in bourgeois countries.

The rout of the Mendelist-Morganist genetics at the historical session of the LAAAS had great international repercussions. The *Mendelist-Morganist pseudoscience, an expression of senile decay and degradation of bourgeois culture,* demonstrated its complete bankruptcy. At its roll call it could summon only the lie which reinforced its reactionary sermon on the immutability of heredity. In the light of the tremendous practical and theoretical achievement of progressive Michurinist science, it became completely evident that Mendelist-Morganist genetics has no right to claim to be a science. It became obvious that it owed its development to the interest taken in it by the forces of the international bourgeoisie.

The complete victory of Lysenko's teaching was marked in our days by the crushing ideological rout of the supporters of the reactionary, antiscientific, Weismannist-Mendelist-Morganist trend in biology. This was one of the victories of socialism, of communism over capitalism.

The "complete victory" came in 1948 when Lysenko's ideas were officially endorsed by Marx's then current vicar on earth, Iosip Vissarionovich Stalin. Actually even earlier on hearing one of Lysenko's speeches, Stalin had exclaimed "Bravo Comrade Lysenko, bravo!"; so his views were well-known in the Soviet Union and they tended to carry great weight. Such was the judgment of the man whose genius in science (as well as in most other fields of human endeavor) has been attested to by uncountable Soviet savants.

Perhaps the most perceptive analysis of the Soviet attitude was expressed in some sardonic stanzas by Einstein (translated from the German by the author):

Wisdom of Dialectical Materialism

By sweat and toil unparalleled
A grain of truth you grasped and held?
Fool! Why suffer to such a degree,
We find truth easily, by Party decree.
And if some doubts you start to mull
You quickly earn a smashed-in skull.
Thus have we produced the key
To a life of blissful harmony.

Stalin was not the only figure with wide-ranging talents in the Marxist world. Announcing the total synthesis of insulin in 1966, Kung Yueh-ting and his collaborators in Shanghai and Peking wrote as follows:

The first total synthesis of a protein [insulin] was accomplished in 1965 in the People's Republic of China. *Holding aloft the great red banner of Mao Tse-tung's thinking and manifesting the superiority of the socialist system, we have achieved, under the correct leadership of our party, the total synthesis of bovine insulin.* . . .Throughout the various stages of our investigation, we followed closely the teaching of Chairman Mao Tse-tung: eliminating superstitions, analyzing contradictions, paying respect to practice, and frequently summing up experiences. . . .The total chemical synthesis of insulin is a piece of work which stems directly from the big leap forward movement.

How much of a big leap this work really was, and in particular, how much had been achieved beyond the accomplishments of Katsoyannis in the United States and Zahn in Germany, is now not so clear. Judging from recent reports from the People's Republic, we may now also entertain some doubts about Mao's teachings.

In all fairness I should point out that political leaders in capitalist countries, whose opinions are generally less murderous, also have a good record of misperceptions. It was the great prime minister of 19th century Great Britain, William Gladstone, who opined that "the present is by no means an age abounding in minds of the first order." That was the judgment of a man who could list among his British contemporaries alone the names of Faraday, Kelvin, Darwin, and Maxwell. This same Gladstone, according to a legendary story, also made the following comment while being escorted through Michael Faraday's laboratory to be shown the revolutionary new discoveries in electromagnetic induction (which became the basis of the entire electrical power industry): "Very interesting, Mr. Faraday, but of what possible use is this?" To this Faraday replied: "Of what possible use is a baby?"

This concludes our non-random walk through the literature of science in search of justifications for administering a few black eyes to scientists of great, or lesser, distinction. It is not difficult to be an infallible Monday-morning quarterback. The crucial challenge, however, is to make a correct judgment ad hoc, not post hoc. We have not been able to discover criteria against which to evaluate each of the many different types of novel ideas arising in all areas of science. These can have different forms of abstraction, different modes of expression, different goals. From the small assembly of cases that we have examined here, it is hazardous to attempt to extract any guides for early recognition of the ultimate value of a new paradigm. Perhaps no such principles exist. Practicing scientists in general must operate on the assumption that generally-accepted fundamental views in their field are valid. They cannot pay serious attention to every challenging conceptual or experimental claim, for they would be inundated with diversionary distractions which, in the overwhelming majority of cases, turn out to be totally without merit. On the whole, it does pay to heed the biblical injunction: "Beware of false prophets."

Perhaps we can reasonably ask, however, that each of us remember the very perceptive epigram of the greatest and most imaginative of all chemists, Antoine Laurent Lavoisier, who said, "The human mind becomes creased into ways of seeing things."

Barber, B. (1961). "Resistance by Scientists to Scientific Discovery". *Science*, Volume 134, pp. 596-602.

Bethe, H. A. (1968). "Energy Production in Stars". *Science*, Volume 161, pp. 541-547.

Dukas, H. and Hoffmann, B. (1979). *Albert Einstein, The Human Side*, Princeton University Press, Princeton, N. J.

Dunn, L. C. (1965). *A Short History of Genetics*, McGraw Hill Book Co., New York.

Hoffman, B. and Dukas, H. (1972). *Albert Einstein, Creator and Rebel*, Viking Press, New York.

Huxley, L. (1901). *Life and Letters of Thomas Henry Huxley*, D. Appleton and Co., New York.

Irvine, W. (1955). *Apes, Angels and Victorians*, McGraw-Hill Book Co., New York.

Jeffrey, E. C. (1925). "Drosophila and the Mutation Hypothesis". *Science*, Volume 62, pp. 3-5.

Kung, Y.-T., and 20 collaborators (1966). "Total Synthesis of Crystalline Insulin". *Scientia Sinica*, Volume 15, pp. 544-545; see also *Science*, Volume 153, pp. 281-283.

Means, J. H. (1964). *James Means and the Problem of Manflight*, Smithsonian Institute, Washington, D. C.

Medvedev, Z. A. (1969). *The Rise and Fall of T. D. Lysenko*, Translated by I. M. Lerner, Columbia University Press, New York.

Morgan, T. H. (1909). "What are the 'Factors' in Mendelian Explanations?", *American Breeder's Association*, Volume 5, pp. 365-368.

Sturtevant, A. H. (1965). *A History of Genetics*, Harper and Row, New York.

Thompson, S. P. (1910). *Life of Lord Kelvin*, Macmillan and Co., London.

Wheeler, W. M. (1923). "The Dry-Rot of Our Academic Biology". *Science*, Volume 57, pp. 61-71.

III
Great Discoveries
Not Mentioned in Textbooks:
N Rays

Practicing cranks in science almost invariably associate their own names with those of giant iconoclasts who turned out to be right — Galileo, Darwin, Einstein. From the writings or lectures of these fanatics, one gains the impression that, in general, history has vindicated the position of scientific heretics in their battles with the scientific establishment. Such a surmise is a natural outcome of the fact that only the iconoclasts who turned out to be right attained a position in history; the myriads who were wrong, or who had nothing substantive to offer, vanished into the limbo of human blunders. It is difficult to find any trace in modern literature of even some of the *grand* illusions of science — spectacular discoveries or novel concepts that excited a large segment of the scientific community, but that turned out to be errors or self-deceptions. As in the arts or religion, a striking idea or an ecstatic personality in science can infect scores of individuals with receptive frames of mind, sometimes even creating a form of mass hysteria. N rays provide a rich illustration of several facets of such a phenomenon in the psychosociology of science.

The period around the turn into the twentieth century was an especially exciting time in modern science. X rays had just been produced. Alpha (α) rays, beta (β) rays and gamma (γ) rays from radioactive materials had just been discovered. Rays were filling the scientific air. Expectations that still others would be discovered were pervasive. Letters of the alphabet between γ and X were all unappropriated, still available.

Such a discovery was indeed announced in 1903 by a distinguished French physicist, Professor René Blondlot, a member of the Academy of Sciences of France (Figure 1). The new rays were christened first "n rays" and then "N rays", to immortalize the city of Nancy in which his university was located.

Blondlot was an accomplished experimentalist in the physics of electromagnetic rays. Among other things, he had studied electric discharges through gases, the type of phenomenon that ultimately led Roentgen (a German) in 1895 to the discovery of X rays. It is not surprising, therefore, that after 1895 Blondlot became intimately involved in X ray research. At the turn of the century, the nature

Figure 1. René-Prosper Blondlot (1849–1930), D.Sc. (Paris) 1881, Professor at the University of Nancy, dressed in the robes of the French Academy of Sciences. The Academy named him to receive three of its prestigious prizes, in recognition of his researches in electromagnetism. (Courtesy of the Academy of Sciences, Paris.)

of the X ray — whether it was a ray of particles or of electromagnetic waves — was a mystery. There have been standard criteria in experimental physics for discriminating between particles and waves in any specific case. Normally beams of particles coming from an electrical source carry an electric charge; such beams passing in a straight line between a positive and a negative electrically-charged plate are bent from their previous direction of motion. Complementarily, a beam of electromagnetic waves, which is not deflected by charged plates, may be bent by a transparent prism or polarized. A polarized wave is one in which the electromagnetic pulsations, normally occurring in all directions in space, are confined to a flat two dimensions so that the moving wave may be represented by a stretched-out, flat ribbon. (Polaroid lenses polarize visible light waves, that is, transmit visible waves pulsating in only one plane.) If the flat plane of this ribbon is up-and-down, the electric pulsations of the wave are vertical. If the ribbon's plane is oriented horizontally, the pulsations are in the horizontal direction. Blondlot argued that if X rays are waves, they might be polarized as they exit from the electric discharge tube in which they are generated. To detect such a polarization, he proposed to use a small electric spark jumping in a straight line between two needle-sharp, pointed wires with a tiny gap (one-twentieth of an inch) between them (see Figure 2). If the line of the

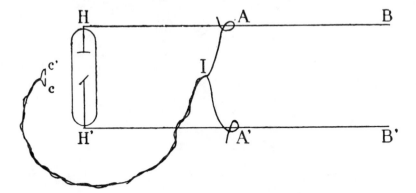

Figure 2. Schematic diagram, published by Blondlot in 1903 (in *Comptes Rendus Academie des Sciences* of France), of apparatus that led him to propose the existence of N rays. The spark gap detector (C–C') is coupled with the discharge-tube circuit (B–H–H'–B). An induction coil, connected at B and B', provided the surge of electricity that generated a pulse of X rays from the discharge tube (H–H') and simultaneously produced a small spark at the spark gap (C–C'). Estimates of the intensity of the spark gap were made visually, first in the orientation shown in the plane of the paper and secondly turned perpendicular to the plane of the paper.

spark was oriented so that it lay in the plane of the ribbon of electric pulsations of the X ray, one might expect the electrical energy of the wave to reinforce that in the spark, and increase its brightness. To his delight, Blondlot found that in a certain orientation with respect to the X-ray generator tube, the spark visibly increased in brightness. His initial excitement was temporarily dampened, however, by the succeeding discovery that the spark gap detector indicated that the radiation from the X-ray tube was bent when it traversed a quartz prism: it had been established unequivocally earlier by photographic detection methods that X rays are *not* bent by a quartz prism.

It was at this point that Blondlot made his inspired conceptual leap. The visible increase in brightness of the spark showed that some wave had impinged on it. If the wave was not an X ray, than it must be some new form of electromagnetic radiation — an N ray.

As an experienced physicist, he knew that he had to rule out effects from traces of ordinary light waves. This he did by enclosing the spark gap detector in a light-tight box (Figure 3). Furthermore, as

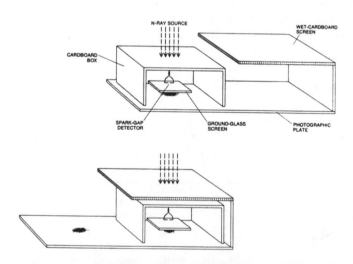

Figure 3. Tangible photographic evidence for the existence of N rays was obtained from experiments of the type illustrated in this diagram, adapted from that described by Blondlot. Again the spark-gap served as the sensitive element for detection. Between it and the recording photographic plate was a ground-glass screen, inserted to diffuse the light from the spark and smear out the image on the plate. The wet-cardboard screen, believed to be opaque to N rays, served as the absorbing filter and could be moved into and out of the path of the beam of N rays.

42

Blondlot recognized, it would be much more convincing if he could discard subjective visual observations and replace them by an objective detection device. For this purpose he designed a clever differential photographic technique. In the apparatus shown in Figure 3, a photographic recording of the spark intensity was obtained, first with the spark gap exposed to N rays, and secondly with the gap protected from the N rays by a suitable blocking screen. If N rays existed, then the blackening on the first photographic frame should be deeper or more extensive than that on the second frame. Furthermore, to magnify such a difference, Blondlot cleverly programmed a series of successive paired exposures with the blocking shutter alternately in and out of the path of the N rays (Figure 3). Some of the photographs that he published (see Figure 4, for example) show unquestionable differences in intensity and are very convincing.

In the course of time, he also developed improved detectors of N rays. In particular, he adapted phosphors, substances that emit a glow of visible color when invisible light impinges upon them. With new techniques he made a broad-ranging study of potential alternative sources of N rays, and of the properties of these novel electromagnetic waves. He was quickly joined in these endeavors by dozens of independent scientists of varying degress of distinction and competence (Table 1).

Figure 4. A typical double frame of paired images, published by Blondlot in 1904 (in *Comptes Rendus Academie des Sciences* of France), showing distinctly more blackening of the plate (right exposure) when the filter was removed from the path of the N rays. In this experiment a Nernst lamp provided the N rays.

Table 1. Some Authors of Publications on N Rays

R. Blondlot	C. Gutton
P. Audollent	G. Jegou
H. Bagard	Lambert
G. Ballet	J. Macé de Lépinay
J. Becquerel	E. Mascart
E. Bichat	P.-L. Mercanton
H. Bordier	E. Meyer
A. Broca	J. Meyer
L. Bull	C. Radzikowski
A. Charpentier	E. Rothé
A. Colson	G. Weiss
	A. Zimmern

Emitters of N rays other than an electric discharge tube were soon discovered. The Welsbach mantle (known as the Auer lamp before Karl Auer was ennobled to Baron Welsbach by the Kaiser), a popular gas lamp for home lighting at the turn of the century, proved to be a rich source of these rays. Curiously enough the workhorse gas burner of scientific laboratories, the Bunsen burner, was impotent. However, the Nernst glower, a lamp with a thin rod of rare-earth oxides heated to incandescence by an electric current (and a popular source of infrared rays), and heated pieces of sheet iron or silver were found to be among the rich sources of N rays. Soon natural sources were added to these artificial ones. Blondlot discovered that the sum emitted N rays. Augustin Charpentier, a professor of medical physics at the University of Nancy, in a report sponsored by d'Arsonval, a famous member of the French Academy of Sciences, described the emission of N rays by the human body, especially by muscles and nerves. Claims were also made that the location of the "genital center" in the spinal medulla could be detected if one added a testicular extract to the sulfide in the phosphor detector. Moreover, devotees of *Star Wars* should be delighted to learn of experiments by Charpentier that also detected N ray emissions after death. Subsequently, a Monsieur Lambert found that even enzymes isolated from tissues emitted N rays.

The materials that are transparent to N rays (Table 2) are as surprising as some of the sources. Almost all of them are opaque to visible light rays. Metals and wood, including paper, were found to

Table 2. Materials Transparent or Opaque to N Rays

Transparent		Opaque	
Black paper	Steel sheet	Water	(Fluorspar)*
Cigarette paper	Silver leaf	Rock Salt	(Sulfur)
Wood	Gold leaf		(Glass)
Aluminum	Mica		
Copper	Paraffin		
Brass	Quartz		
Tinfoil			

*Items in parentheses are partly opaque to N rays.

transmit N rays efficiently. In fact, Blondlot fabricated the lenses and prisms he needed to focus and bend N rays, out of aluminum, a relatively novel and scarce metal in his day. In contrast, water and rock salt were reported to be opaque to N rays and hence could serve as shutters to obstruct passage of these waves.

Blondlot also constructed more sophisticated equipment to study the bending or refraction of N rays quantitatively (Figure 5). With this apparatus he demonstrated that independent observers found the refracted N rays at precisely the same angle of refraction, that is, at the same position of the ruling engine (Table 3). His summary comment at the end of the article in which these data were published reads "given the results [in this Table] I let everyone reach his own conclusion."

With this refractometer Blondlot was also able to measure the wavelength of N rays, by comparison of their refractive behavior with that of yellow light. From the information he provided we can calculate that N rays had wavelengths of about 100 Å.

The very wide interest in and prominent status of N rays in experimental physics in 1904 was manifested by the large number of papers published on the subject. Figure 6 shows a reproduction of one index of the *Comptes Rendus,* the major publication of the Academy of Sciences in Paris. It is obvious that the number of papers on N rays far exceeded that on X rays.

Even at the beginning of this century there were many prestigious awards to bestow on distinguished scientists. In 1904, the French Academy honored, and enriched, Blondlot by presenting the Prix Leconte to him. The citation (Figure 7) reads that the award (of

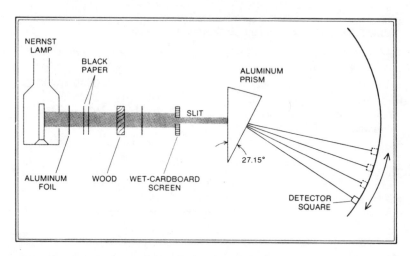

Figure 5. Schematic diagram of equipment, described by Blondlot in 1904–1905, used to demonstrate refraction of N rays. Radiation from the Nernst lamp was passed through sheets of aluminum, black paper and wood to filter out all forms of radiation except N rays. A screen of wet cardboard with a vertical slit 3 millimeters wide was placed in the optical path to define a moderately sharp beam that impinged on the aluminum prism. According to Blondlot the prism refracted (that is, bent) the N rays, and also spread them out into a spectrum of lines, presumably of different wavelengths. To detect the deviated rays Blondlot used a small piece of cardboard, with a phosphorescent strip painted down the middle, mounted on a curved steel support with a calibrated screw from a ruling engine. An increase in luminosity of the phosphorescent strip signaled the presence of a refracted N ray. This position was recorded precisely. Blondlot and his colleagues recorded such signals at several very reproducible positions.

50,000 francs) was "pour l'ensemble de ses travaux" (for the total body of his work) and mentions the "new rays" circumspectly, only at the end of a detailed three-page recitation of Blondlot's achievements. Nevertheless, the concluding sentence affirms the confidence of the committee in the experimenter and extends its expression of support in his difficult new researches. French scientific gossip of that time alleges that the citation originally pointed only to the discovery of N rays, but that caution among committee members subsequently prevailed. In some respects the public citation reminds one of that of the Nobel Committee in its 1922 award to Einstein, when it referred explicitly to his "discovery of the law of the photo-electric effect" but alluded only indirectly to relativity with the phrase "for his services to theoretical physics."

Table 3. Comparison of Readings of Refraction of N Rays
Made by Four Independent Experimenters*

Blondlot	Gutton	Virtz	Mascart	Mean
382.4	-	381	383.4	382.4
-	387.2	386.9	387	387.03
391.5	393	392	391	391.9
398.4	399	398.2	397	398.15

* E. Mascart was a professor of physics at the College de France in Paris as well as on the editorial staff of *Comptes Rendus* of the Academy; C. Gutton became professor at Nancy in 1906; L. Virtz was Blondlot's mechanic and professional assistant.

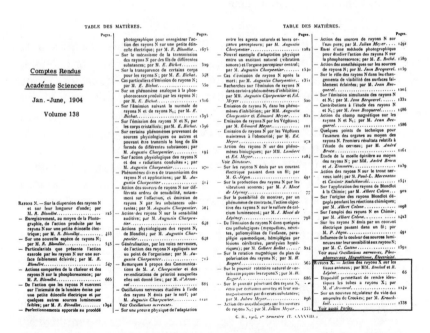

Figure 6. Extract from the index of the proceedings of the French Academy of Sciences showing the listings of all the papers published on N rays during the first half of 1904 and the listings of those published on X rays. The ratio is 53 to 3.

PRIX LECONTE.

(Commissaires : MM. Mascart, Troost, Darboux, Berthelot, Maurice Levy, H. Becquerel, Bouchard, Moissan, Janssen, de Lapparent; Poincaré, rapporteur.)

La Commission nommée pour décerner le prix Leconte en 1904 a porté son choix sur M. René Blondlot, Correspondant de l'Académie des Sciences, Professeur à la Faculté des Sciences de Nancy, pour l'ensemble de ses travaux.

Figure 7. Formal announcement of the French Academy of Sciences in 1904 (*Comptes Rendus,* Vol. 139, pp. 1120–1122) that M. René Blondlot had been selected as the awardee of the Prix Leconte. The runner-up in the selection process was Pierre Curie. Curie and Blondlot had published jointly in 1889. Note the many distinguished scientists who were members of the award committee. The monetary value of the Prix Leconte was 50,000 francs, at that time equivalent to $12,000, a sum that was 4–10 times the annual salary of an American or French professor around 1900. Blondlot had also received two earlier prizes from the Academy, one in 1893 and the second in 1899. These were given in recognition of his researches on electric waves. His accomplishments in that area had been highly praised by H. Poincaré and J. J. Thomson, among others.

Like previous great discoverers, Blondlot and his disciples were faced with claims of priority from others. In the spring of 1903, Monsieur Gustave Le Bon, a widely-known French savant, wrote that seven years earlier he had discovered emanations capable of traversing metals. In December 1903, a Monsieur P. Audollent presented a petition to the French Academy claiming priority over A. Charpentier in the discovery of N-ray emissions from natural bodies. In January 1904, Monsieur Carl Huter, a spiritualist, entered a similar claim. In the spring, the Academy solemnly announced its judgment, in a report by d'Arsonval, that priority belonged to Charpentier.

Of these three challengers, M. Le Bon is the most intriguing. He has recently become the darling of a small splinter conservative group in this country that has tried to revive Le Bon's once formidable reputation in French intellectual circles. His book *The Psychology of the Crowd* has been called "the most influential book ever written in social psychology", and this Renaissance-type "homo multiplex" is also purported to have anticipated Einstein in conceiving the theory of relativity ("in a non-mathematical form") and to have foreseen nuclear processes. His modern disciples claim that in the scientific establishment there has been a conspiracy of silence about Le Bon and that scientists have tried to destroy him because he expounded a conservative, anti-leftist social philosophy.

Not being able to judge Le Bon's contributions to sociology, I thought it might be interesting to read some of his articles on physical sciences in the *Revue Scientifique,* such as the one entitled "La Materialisation de l'énergie." I was particularly intrigued as to what a "non-mathematical" theory of relativity might be.

The essays in the *Revue* turn out to be just speculative ramblings with little substance and no enduring value. For example, Le Bon claimed that when matter "dematerializes" or "dissociates," it passes through various "phases" ending up ultimately as "ether"; that electricity constitutes a general form of the dematerialization of energy; that the products of the dematerialization of atoms are constituted of substances intermediate in properties between matter and ether. In the course of these ramblings one also finds clauses stating that matter can be converted into energy. To interpret such clauses as evidence that Le Bon anticipated the theory of relativity shows a total ignorance of the essence of Einstein's approach, let alone of his theoretical constructs. (It is of interest that the Nazis made a similar claim based on similar reasoning, that their "pure Aryan," F. Hasenöhrl had anticipated Einstein. Hasenöhrl was at least a respected theoretical physicist, a contemporary of young Einstein, but Hasenöhrl never in his wildest dreams felt that he had anticipated special relativity.) After reading Le Bon's writings in the *Revue,* one is not surprised that he could claim that he had discovered Blondlot's rays seven years before Blondlot. Perhaps silence is the more compassionate treatment of Le Bon.

Research in N rays flourished — for about a year and a half. The number of scientific papers published rose explosively, as in fact is characteristic in most exciting new research areas (Figure 8). As an illustration, one can point to the very novel enzymes, restriction endonucleases, for which a Nobel Prize was given in 1978. After the discovery of these remarkable enzymes in the late sixties, the number of scientific research papers rose very gradually for a few years and then very steeply from the early seventies onward, as molecular biologists verified the existence of these naturally-occurring catalysts and extended enormously the range of their scientific applicability. The quantitative historiography of N rays shows a similar publication profile initially (Figure 8). However, a catastrophic, irreversible collapse in the pattern, atypical of great discoveries, occurred in 1905. What happened?

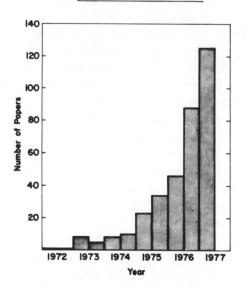

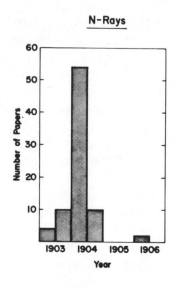

Figure 8. Profile of publications as a function of time. Those in the area of restriction endonucleases rose almost logarithmically in the 1970s. It is difficult to search thoroughly for publications after about 1977 because the use of these enzymes became so pervasive that the titles of papers no longer reveal that these materials provided an important tool for the research described. It is likely that the curve is still rising steeply since restriction endonucleases provide crucial devices in multitudinous genetic engineering projects. The profile of publications shown for N rays is based on entries in *Comptes Rendus.* Its height would be greater if all of the approximately 200 papers published everywhere had been included, but the shape (similar to what mathematical theorists call a "delta function") would not alter.

What shattered the edifice of N rays was a scientific note published in 1904 in *Nature,* a British scientific weekly, by the famous American physicist, R. W. Wood.

Robert Williams Wood was an internationally renowned scientist of the first half of this century. Contrary to popular images of a scientist, Wood was outgoing, ebullient in personality, perceptive and curious about everything in his physical and social environment, creative in many genres. In his time and locale, he was as well known to the public as that other sage of Baltimore, H. L. Mencken. An example of Wood's many talents is his little book *How to Tell the Birds from the Flowers* (Figure 9), a collection of humorous jingles attached to his individualistic drawings of animals and plants. It was intended as a parody of nature books, "a revised manual of flornithology for beginners." Its flavor can be savored in two of the

How To Tell The Birds From The Flowers
And Other Wood-cuts.

A Revised Manual of Flornithology.for Beginners.

Verses and Illustrations
By Robert Williams Wood.

Figure 9. Title page of R. W. Wood's book *How to Tell the Birds from the Flowers*, first published in 1917 and reprinted in 1959. (Courtesy of Dover Publications, Inc.)

doggerels (Figure 10), one contrasting parrot with carrot, the other elk with whelk.

As his biographer (William Seabrook) pointed out, Wood was a small boy who became a famous man but never grew up. His childhood pranks, such as planting pounds of explosive mixtures of potassium chlorate and sulfur in a public street, were succeeded by adolescent hoaxes, such as planting a story in a Chicago newspaper describing his discovery and retrieval of an extraterrestrial flying object that had hit the earth the preceding evening, "a phenomenon unparalleled in the annals of astronomical science, which throws light upon the question of habitability of other planets." The steel-blue ball was engraved with pictures and writings, resembling those of shorthand; chemical analysis purportedly showed that it was made of a totally new element and hence could not have been manufact-

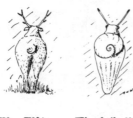

The Elk. The Whelk.

A roar of welkome through the welkin
Is certain proof you'll find the Elk in;
But if you listen to the shell,
In which the Whelk is said to dwell,
And hear a roar, beyond a doubt
It indicates the Whelk is out.

The Parrot. The Carrot.

The Parrot and the Carrot one may
easily confound,
They're very much alike in looks
and similar in sound,
We recognize the Parrot by his
clear articulation,
For Carrots are unable to engage
in conversation.

Figure 10. Two selections from R. W. Wood's creation *How to Tell the Birds from the Flowers*. (Courtesy of Dover Publications, Inc.)

ured on earth. Present day UFO enthusiasts would classify this incident as a CEIII, close encounter, type III. Later, Wood combined his extraordinary talents in optics with his interest in extraterrestrial objects by revealing a photograph that must be one of the very first ever snapped of a UFO, taking off from the moon (Figure 11).

As a student Wood went out of his way to irritate authorities. During his graduate training in Berlin, at the end of the last century, he once unobtrusively bought a second class, green ticket for the city rail lines and slipped it into an inner pocket, and then conspicuously purchased a third class, yellow ticket that he flashed to the platform guard as he hurriedly rushed into the second class car of a waiting train. The guard rushed in after him and began to shout in German at Wood, who pretended not to understand. When the guard ordered Wood to get off at the very next station, Wood protested in broken German that his destination was further on. The guard in a rage finally managed to get Wood off the train, at Wood's destination, and to call police to have Wood arrested. At this point Wood pulled out his green second-class ticket and in fluent German insisted that the guard must be color blind.

Later, the mature man, now Professor Wood of Johns Hopkins University (Figure 12), walking along the puddle-laden street of a heavily-populated poor neighborhood of Baltimore, would ostentatiously spit a wad of saliva into the puddle while unobtrusively

MOONSHINE: One of the photographs Wood faked for illustrating *The Moon-Maker*, the pseudo-scientific "thriller" on which he and Arthur Train collaborated. The "flying ring" is taking off from the surface of the moon.

Figure 11. A photograph of an extraterrestrial object, known today as a UFO, taking off from the moon. Wood was recognized as a genius in physical optics and hence it was credible that he had constructed a telescope capable of providing the detail shown in this reproduction. (From William Seabrook's biography *Doctor Wood*, with permission of Harcourt, Brace, Jovanovich and Co.)

flipping in also a pellet of metallic sodium, which on contact explodes into a geyser of sparks, yellow fire, and smoke. His demonstration of satanic powers generated pandemonium in such a locale in those days.

His style even in his scientific work, was that of a master showman. Early in his university career, he designed and personally constructed the largest and best spectroscope in the world. The core of the instrument was a forty-two foot long optical tube, about six inches in diameter. Since it was large, the tube was mounted in a barn on his property at East Hampton (Figure 13). Construction of the auxiliary equipment took most of a winter and spring. By

Figure 12. Robert Williams Wood (1868–1955) as he appeared in 1901 at the time he was appointed professor of physics at Johns Hopkins University, where he succeeded Henry Rowland. He never earned a doctoral degree, but at the time of his appointment at Hopkins he had published thirty papers. Ultimately he received six honorary degrees, was elected a member of the National Academy of Sciences (USA) and was one of the very few foreigners elected a foreign fellow of the Royal Society of London. Some of the other company he mingled with are shown in Figure 14. Wood continued to publish research papers up to age eighty-one. His last publication, in 1949, described some very remarkable features of the crystals of protocatechuic acid. His observations and qualitative interpretations of them were fully confirmed and delineated in detail in a modern X-ray study of these crystals in 1983 by F. H. Herbstein and I. Agmon. (Photograph from William Seabrook's biography *Dr. Wood,* with permission of Harcourt, Brace Jovanovich and Co.)

June, Wood found the tube blocked with spider webs. As described in his published technical report in the *Philosophical Magazine* of 1912:

> Some trouble was given by spiders, which built their webs at intervals along the tube, a difficulty which I surmounted by sending our pussy-cat through it.

Figure 13. Photograph of the forty-two-foot spectroscope constructed by Wood on his Long Island farm around 1910. The optical tube was so long that it projected through the wall of the barn. At that time this instrument had the highest resolving power of any spectroscope in the world. Wood's spectroscopic data provided basic information needed by many theorists, including Niels Bohr. (Reproduced from Wood's article in the *Philosophical Magazine* in 1912.)

Figure 14. R. W. Wood, M. Planck and A. Einstein on an occasion (about 1930) when Wood gave a lecture in Berlin. (From William Seabrook's biography *Dr. Wood*, with permission of Harcourt, Brace, Jovanovich and Co.)

A flair for showmanship also dominated his merciless treatment of cranks and quacks. One of the early purveyors of "death rays," an Englishman named Grindell Matthews, was described by Wood after examination of the ray generator, as "self-deluded or not... [comparable to] promoters who try to sell Brooklyn bridges to innocent bystanders." Wood was also the nemesis of mediums. When a famous Italian medium, Eusapia Palladino, visited the United States before World War I, Wood was invited to join an investigatory committee sponsored by the then *Scientific American.* For those who claimed to be able to communicate with the dead, Wood had only contempt. In responding to an invitation from one such spiritualist to communicate with any recently departed friend, Wood took a fiendish delight in proposing subtle mathematical questions to the ghost of Lord Rayleigh, a very great British theoretical physicist who dealt with some of the most abstruse aspects of electromagnetic phenomena. On the other hand, Wood admired performers such as Palladino, who was a skilled physical medium, rather than a psychic or spiritualist one. Among other things, she could make objects (such as a tamborine) move from a curtained cabinet behind her to the table on which her hands were resting. As Wood said, "I convinced myself very early in the sittings that all the phenomena were fraud." Nevertheless, he was unable, despite his ingenuity, to trap her. After one of her sittings, he installed an X-ray lamp on one side of the cabinet and a fluorescing detector screen on the other so that he would be able to see any mechanical or human arm that reached into the curtained cabinet. However, Mme. Palladino was very canny; she feigned illness and refused to appear at any additional sittings.

As a spectroscopist — an expert in electromagnetic rays (Figure 14) — Wood was immediately excited by Blondlot's discoveries. As is customary when someone announces a novel phenomenon in science, Wood set about to try to reproduce the striking experiments. He assembled the necessary equipment and, as he said to his biographer in his scornful manner, "I attempted to repeat his [Blondlot's] observations, but failed to confirm them after wasting a whole morning." Other highly respected physicists, including Lord Kelvin and Sir William Crookes in Great Britain and Heinrich Rubens and Otto Lummer in Germany, had been equally unsuccessful in their attempts to discover any evidence of the existence of N rays. Under their urging, Wood agreed to visit Blondlot's laboratory, where, as Wood states, "the apparently peculiar conditions necessary for the manifestation of this most elusive form of radiation appear to exist." There he was cordially received, at least at the beginning of his visit, and was treated to a series of experimental demonstrations of the properties of N rays. Wood feigned to be unable to converse in French (so that he could eavesdrop on conversations between Blondlot and his assistant) and communicated with Blondlot in German. An account of his experiences in the Blondlot laboratory was published by Wood in the September 29, 1904 issue of *Nature,* which has always had a wide circulation in the scientific community internationally. Wood did not identify Blondlot in this article, but in conversations with his biographer, Wood disclosed that all the experiences described in his letter to *Nature* did take place in Nancy. After a short introduction, the article reads as follows:

The first experiment which it was my privilege to witness was the supposed brightening of a small electric spark when the n-rays were concentrated on it by means of an aluminum lens. The spark was placed behind a small screen of ground glass to diffuse the light, the luminosity of which was supposed to change when the hand was interposed between the spark and the source of the n-rays.

It was claimed that this was most distinctly noticeable, yet I was unable to detect the slightest change. This was explained as due to a lack of sensitiveness of my eyes, and to test the matter I suggested that the attempt be made to announce the exact moments at which I introduced my hand into the path of the rays, by observing the screen. In no case was a correct answer given, the screen being announced as bright and dark in alternation when my hand was held motionless in the path of the rays, while the fluctuations observed when I moved my hand bore no relation whatever to its movements.

Blondlot, a skilled physicist, had been aware even before Wood's visit that judgments of light intensity by eye are notoriously unreliable and, as mentioned earlier, had already developed a photographic apparatus to record increases in spark brilliance upon exposure to N rays. Wood reported:

I was shown a number of photographs which showed the brightening of the image, and a plate was exposed in my presence, but they were made, it seems to me, under conditions which admit of many sources of error. In the first place, the brilliancy of the spark fluctuates all the time by an amount which I estimated at 25 percent, which alone would make accurate work impossible. Secondly, the two images (with N rays and without) are built of "installment exposures" of five seconds each, the plate holder being shifted back and forth by hand every five seconds. It appears to me that it is quite possible that the difference in the brilliancy of the images is due to a cumulative favouring of the exposure of one of the images, which may be quite unconscious, but may be governed by the previous knowledge of the disposition of the apparatus. The claim is made that all accidents of this nature are made impossible by changing the conditions, i.e., by shifting the positions of the screens; but it must be remembered that the experimenter is aware of the change, and may be unconsciously influenced to hold the plate holder a fraction of a second longer on one side than on the other.... [T]o be sure the photographs are offered as an objective proof of the effect of the rays upon the luminosity of the spark. The spark, however, varies greatly in intensity from moment to moment, and the manner in which the exposures are made appears to me to be especially favourable to the introduction of errors in the total time of exposure which each image receives. I am unwilling also to believe that a change of intensity which the average eye cannot detect when the N rays are flashed "on" and "off" will be brought out as distinctly in photographs, as is the case on the plates exhibited.

Experiments could be easily devised which would settle the matter beyond all doubt; for example, the following: — Let two screens be prepared, one composed of two sheets of thin aluminum with a few sheets of wet paper between, the whole hermetically sealed with wax along the edges. The other screen to be exactly similar, containing, however, dry paper.

Let a dozen or more photographs be taken with the two screens, the person exposing the plates being ignorant of which screen was used in each case. One of the screens being opaque to the N rays, the other transparent, the resulting photographs would tell the story. Two observers would be required, one to change the screens and keep a record of the one used in each case, the other to expose the plates.

The same screen should be used for two or three successive exposures, in one or more cases, and it should be made impossible for the person exposing the plates to know in any way whether a change had been made or not. I feel very sure that a day spent on some such experiment as this would show that the variations in the density on the photographic plate had no connection with the screen used.

In essence Wood conceded that the photographic plate experiments supported Blondlot's contention, but as a consummate skeptic or critical scientist, depending on one's position, Wood rationalized the observations to conform with his conviction before arrival in France. At this stage, one round must be awarded to each contestant.

I was next shown the experiment of the deviation of the rays by an aluminum prism [see Figure 5]. The aluminum lens was removed, and a screen of wet cardboard furnished with a vertical slit about 3 mm. wide put in its place. In front of the slit stood the prism, which was supposed not only to bend the sheet of rays, but to spread it out into a spectrum. The positions of the deviated rays were located by a narrow vertical line of phosphorescent paint perhaps 0.5 mm. wide, on a piece of dry cardboard, which was moved along by means of a small dividing engine. It was claimed that a movement of the screw corresponding to a motion of less than 0.1 of a millimetre was sufficient to cause the phosphorescent line to change in luminosity when it was moved across the N ray spectrum, and this with a slit 2 or 3 mm. wide. I expressed surprise that a ray bundle 3 mm. in width could be split up into a spectrum with maxima and minima less than 0.1 of a millimetre apart, and was told that this was one of the inexplicable and astounding properties of the rays. I was unable to see any change whatever in the brilliancy of the phosphorescent line as I moved it along, and I subsequently found that the removal of the prism (we were in a dark room) did not seem to interfere in any way with the location of the maxima and minima in the deviated (!) ray bundle.

This published description of the surreptitious removal of the aluminum prism, so similar to Wood's brutal practical jokes, knocked the props from under N rays. Of course, we have only Wood's word that he removed the prism.

Having delivered his knockout blow, Wood did not stop, however, but stomped mercilessly on his prostrate opponent by continuing his report with a description of claims that strained credulity beyond the snapping point. In early 1904, Charpentier had reported that increases in N ray emission accompanying nerve and muscle motor activity provided a basis for novel, successful clinical explorations, for example, to outline the area of the heart or to localize the site of speech articulation in the brain. These successes prompted him to look into the complementary question, the effects of N rays on various physiological activities, and indeed he discovered that N rays sharpen the senses, e.g., of smell or of hearing. Prompted by these findings, or having discovered them independently, Blondlot and his colleagues by late 1904 had designed a series of experiments to demonstrate increased sensitivity of the sense of sight in the presence of a source of N rays.

I was next shown an experiment of a different nature. A small screen on which a number of circles had been painted with luminous paint was placed on the table in the dark room. The approach of a large steel file was supposed to alter the appearance of the spots, causing them to appear more distinct and less nebulous. I could see no change myself, though the phenomenon was described as open to no question, the change being very marked. Holding the file behind my back, I moved my arm slightly towards and away from the screen. The same changes were described by my colleague. A clock face in a dimly lighted room was believed to become much more distinct and brighter when the file was held before the eyes, owing to some peculiar effect which the rays emitted by the file exerted on the retina. I was unable to see the slightest change, though my colleague said that he could see the hands distinctly when he held the file near his eyes, while they were quite invisible when the file was removed. The room was dimly lighted by a gas jet turned down low, which made blank experiments impossible. My colleague could see the change just as well when I held the file before his face, and the substitution of a piece of wood of the same size and shape as the file in no way interfered with the experiment. The substitution was of course unknown to the observer.

Wood summed up his visit with the comment:

After spending three hours or more in witnessing various experiments, I am not only unable to report a single observation which appeared to indicate the existence of the rays, but left with a very firm conviction that the few experimenters who have obtained positive results have been in some way deluded.

Nevertheless, the defenders of N rays were not left speechless by this attack. After all, some twenty different researchers (Table 1), including well-known names such as A. Charpentier and Jean Becquerel (son of Henri Becquerel, discoverer of radioactivity) had confirmed and extended Blondlot's observations. Furthermore, many others had seen the phenomena at the University of Nancy, or had attended demonstrations presented in Paris by Nancy scientists. Blondlot wrote in one of his articles:

Several eminent physicists, who have been good enough to visit my laboratory, have witnessed them [the photographic detection experiments]. Of these forty experiments, one was unsuccessful.... I believe this failure, unique, be it noted, to be due to insufficient regulation of the spark, which undoubtedly was not sensitive.

In a letter to the *Scientific American,* Blondlot claimed that it was also difficult for him to converse in German (Wood having claimed he could not speak French) and that this had led to misunderstandings during Wood's visit.

In addition, Blondlot went to great pains to counter Wood's criticisms of the photographic installment technique. In a 1905 article, Blondlot described procedures for measuring exposure times automatically and for checking the stability of the spark detector by means of a telephone receiver inserted in its power supply. Furthermore, an extra "time bonus" was allotted to the control experiments, the photographic recording in the absence of the N ray source; exposure times were ½ to 1½ seconds longer. Numerous other precautions and checks were introduced. A large number of results were then presented showing striking differences in photographic image intensities obtained with and without N rays. The editors of the *Scientific American* concluded that "these experiments seem to be free from any objection."

In 1905, Blondlot also published a set of expanded instructions on how to observe the action of N rays:

> It is *indispensable* in these experiments to avoid all strain on the eye, all effort, whether visual or for eye accommodation, and in no way to try to *fix* the eye upon the luminous source, whose variations in glow one wishes to ascertain. On the contrary, one must, so to say, see the source without looking at it, and even direct one's glance vaguely in a neighbouring direction. The observer must play an absolutely passive part, under penalty of seeing nothing. Silence should be observed as much as possible. Any smoke, and especially tobacco smoke, must be carefully avoided, as being liable to perturb or even entirely to mask the effect of the N rays. When viewing the screen or luminous object, no attempt at eye-accommodation should be made. In fact, the observer should accustom himself to look at the screen just as a painter, and in particular an "impressionist" painter, would look at a landscape. To attain this requires some practice, and is not an easy task. Some people, in fact, never succeed.

To bolster his stand on "sensitivity" Blondlot quoted some comments of Helmholtz on photometric measurements, made in the latter's *Handbook of Physiological Optics:*

> This requires much practice and consequently many facts of this nature cannot even be detected without extensive preliminary practice in physiological optics, not even by people who are well-trained in other observations. Thus in many areas we are limited to the observations of just a very few individuals.

That in essence was Blondlot's stand. Of course, such a position is also taken by adherents to the validity of extrasensory perception.

Thus the focus repeatedly was on "sensitivity," but this time the spotlight was turned on the observer. Sensitivity, according to the adherents to N rays, was the central issue.

By 1905, only Frenchmen remained in the N ray camp. Some French scientists charged foreign defamers with scurrilous chauvinism. Others suggested that only the Latin races possessed the sharp sensitivities, intellectual as well as sensory, needed to detect N rays. Anglo-Saxon perceptivity was allegedly dulled by continual exposure to fog, and that of the Teutons was blunted by constant ingestion of beer. Such conclusions, of course, ignored the presence of celebrated Frenchmen, such as Jean Perrin (Nobel laureate in 1926) in the anti-N-ray camp.

Actually the severest and most relentless open critics of N rays were Frenchmen. To the French skeptics belongs as much credit as to R. W. Wood for the demise of these radiations. Originally the announcement by Blondlot of the discovery of N rays triggered much excitement. Everywhere in France, physicists, chemists, physiologists, psychologists, etc. jumped on the bandwagon to exploit this fertile research problem. The vast majority of these research workers, including the famous physicists Pierre Curie, Aimé Cotton, Paul Langevin, and Jean Perrin, failed in their attempts to confirm manifestations of N rays. In consequence, they began to look at the published successful observations more critically. Careful experiments, with precautions analogous to those suggested by Wood, invariably failed to detect N rays. Communication among these unsuccessful experimenters soon revealed that there was a curious geographic localization of positive results — Nancy. One exception was the laboratory of Jean Becquerel in Paris. However, when he reported that with chloroform he could anesthetize a piece of metal so that it would not emit N rays, his credibility became suspect.

A drive to resolve the question of N rays was spearheaded by the editors of the *Revue Scientifique,* a popular French scientific journal. They were particularly sensitive to the possible embarrassment of French science and wanted French scientists as Frenchmen to take a courageous critical stand. Articles appeared in successive issues, with titles such as "Les rayons N existent-ils?", in which experiments purporting to establish the authenticity of N rays were criticized severely. Reports were given of the proceedings of international scientific meetings in which English, Belgian, Swiss, Russian, German, etc. and French scientists had voiced their failures to verify reported physiological manifestations of N rays. The *Revue* quoted an eminent Belgian neurophysiologist, A. Waller, who proposed sarcastically that the rays of Nancy be designated "suggestion rays," since the Nancy school was a celebrated center of studies on suggestion, an allusion to Charpentier's earlier work on hypnotism.

The appellation "suggestion rays" is more appropriate than Waller may have realized. Had the Nancy observations been made in the 1880s, ten years before the discovery of radioactivity and of X rays, they would have been subjected to immediate critical analysis, for such rays would have been without precedent. Even the discoverer would have been self-critically cautious. In the early 1900s, however, rays were permeating the air of the frontiers of physics. Even a distinguished physicist was psychologically prepared to stumble into the discovery of a new type of ray. A host of other scientists were also psychologically receptive as soon as the suggestion was explicitly brought to their attention. Observations, especially those on the borderline of reliability, can be fitted into alternative conceptual frameworks, including that of sensory illusion. Choices usually depend on the reproducibility of observations. The convinced personality is unlikely to be capable of an uncommitted analysis.

The *Revue Scientifique* made a valiant effort to provoke Blondlot into a definitive test of the authenticity of N rays. Following suggestions proposed by several physicists, particularly one named Debierne, the *Revue* in 1904 proposed the following experiment. Two identical small wooden boxes should be constructed. In one of these should be placed a small piece of lead, in the other a piece of tempered steel (one of the alleged sources of N rays). The contents of the boxes should be adjusted so that nothing inside moves around and so that their weights are identical. The boxes should then be numbered. Once the boxes are closed and sealed, it should be impossible for anyone to tell which contains the source of N rays. The two boxes should then be submitted to Blondlot to determine with his spark detector or his phosphor screen which one is emitting N rays, that is, which contains the tempered steel. Blondlot did not respond for a long time, but finally in 1906, he wrote:

> Permit me to decline totally your proposal to cooperate in this over-simple experiment; the phenomena are much too delicate for that. Let each one form his personal opinion about N rays, either from his own experiments or from those of others in whom he has confidence.

In a dramatic novel, or movie, the script would undoubtedly have called for Blondlot to commit suicide or to be confined to an insane asylum. Quite the contrary happened, however. Blondlot never gave up his conviction that N rays were real. True, he stopped publishing (although in 1909 he did issue a second edition of his book *Thermodynamics*) and in 1909 he retired from his professorship. Nevertheless, he doggedly pursued his investigations of N rays and submitted sealed notes on his observations to the French Academy. Some of his colleagues at the University of Nancy supported him to the time of his death.

Looking back we can recognize possible personal and sociological pressures that contributed to Blondlot's original misstep and to his persistence in aberrant views even when it became crystal clear that they were untenable. The humiliating military defeat by the Germans in 1870 wounded French pride deeply. Consciously or subconsciously French scientists sensed that their stature was declining, and even more irritating, that the reputation of German science was rising. There was enormous social pressure to outdo the Germans and to strike back at them whenever possible. Pasteur's response was to turn back in 1871 an honorary degree of Doctor of Medicine he had received from the University of Bonn in 1868, with the comment: "The sight of this parchment is odious to me.... [I] ask you to erase my name from the Archives of your faculty." Years later, in response to an inquiry about whether he would accept the Order of Merit from the Kaiser, Pasteur made it abundantly clear that the answer was no.

For Blondlot, a sense of competition with and animosity toward the Germans may have been even stronger. He was a native of Nancy and a member of an elite family in the province (his father had been a member of the medical faculty at the University). Nancy had been historically the capital of Lorraine, but the Germans in 1871 annexed a large part of this province (as well as almost all of Alsace) . Thus Nanciennes had been most painfully hurt, politically and psychologically by the Germans. Furthermore, Nancy was then practically at the German border. Blondlot had seen all of this personally, for he was twenty-one at the time of the Franco-Prussian War.

In addition Blondlot must have been personally chagrined to have missed making the discovery of X rays himself. After all, he had devoted much of his research effort since the early 1880s to electrical discharges and was a recognized expert in the field, having been awarded two of the prestigious prizes of the French Academy. If he had been just a little more perceptive or imaginative, he himself might have discovered X rays. At that time X rays excited the public imagination and were a more spectacular discovery than radioactivity, whose ultimate impact was not yet dreamed of. Thus the first Nobel Prize in physics was awarded to Roentgen; Becquerel and the Curies were so honored only later. Blondlot, and France, thus came very close to getting the credit for the discovery of X rays, and getting the chance of mortifying the hated Germans.

Consequently, Blondlot was in a prepared frame of mind when his first, certainly honest, observations (subject to the subtle experimental effects of ionization of air and heating caused by X rays and the discharge apparatus and to the deceptions that can be created by peripheral vision) could be interpreted as manifestations of a novel form of radiation. Here was a stroke of scientific good luck that would allow Blondlot the Frenchman to surpass Roentgen the German, to vindicate French science and to achieve exceptional fame and glory himself. Once enamored with his discovery, Blondlot bent further observations to conform with his subconscious wishes. He became increasingly enmeshed in a web of incredibilities. Because of his personal and professional pride as well as social pressure, it became almost impossible for him to concede that he had made an error. With their inferiority complex, French scientists were constantly sensitive to the worry articulated by J. A. Le Bel (discoverer, simultaneously with van't Hoff, of the basis of optical activity):

> What a spectacle for French science that one of its distinguished savants determines the position of lines of the spectrum while the prism sits in the pocket of his American colleague.

Thus Blondlot never publicly or probably even consciously changed his stand that N rays were real. As he said after numerous attacks, "I affirm most positively that the phenomena of N rays have for me the same certainty that other physical phenomena have."

When was the decision made that N rays are a delusion? Never. There is no authoritarian hierarchy in science. Science has no vicar on earth to reveal doctrine, no central committee to proclaim the party line. Actual histories of novel discoveries, except when religious or political authorities have intervened, have followed a pattern best described by an aphorism attributed to the founder of the mathematical theory of electromagnetic waves, James Clark Maxwell. In his introductory lecture on light, Maxwell is said to have remarked:

> There are two theories of the nature of light, the corpuscle theory and the wave theory; we used to believe in the corpuscle theory; now we believe in the wave theory because all who believed in the corpuscle theory have died.

René Prosper Blondlot died, at Nancy, in 1930.

Anonymous (1907). "Science and Accuracy". *The Spectator,* Volume 99, issue of October 12, page 520.

Blondlot, R. (1903-1905). "On the Polarization of X Rays", *Comptes Rendus Academie des Sciences,* Volume 136, pages 284-288; "A New Species of Light", *ibid.,* pages 735-738; and many other papers.

Blondlot, R. (1904). Explanations and Statements Concerning N-Rays: A Reply to Prof. Wood". *Scientific American Supplement,* Volume 58, issue of December 17, p. 24211.

Blondlot, R. (1905). *"N" Rays,* translated by J. Garcin, Longmans, Green and Co., London.

Blondlot, R. (1909). *Introduction à L'Etude de la Thermodynamique,* Deuxieme Edition, Gauthier-Villars, Paris.

Editors (1904). "Do N Rays Exist?", *Revue Scientifique,* Series 5, Volume II, Number 18, Issue of October 29, pp. 545-552; "The Solution to the Problem of the Existence of N Rays", Number 23, Issue of December 3, pp. 705-709.

Gillispie, C. C., Editor (1970-1980). *Dictionary of Scientific Biography,* Charles Scribner's Sons, New York.

Klotz, I. M. (1980). "The N-Ray Affair", *Scientific American,* Volume 242, May issue, pp. 168-175.

Le Bon, G. (1904). "The Materialization of Energy", *Revue Scientifique,* Series 5, Volume II, Number 16, Issue of October 15; "The Dematerialization of Matter", Number 20, Issue of November 12.

Nye, M. J. (1980). "N-Rays: An Episode in the History and Psychology of Science", *Historical Studies in the Physical Sciences,* Vol. 11, pp. 125-156.

Pieron, H. (1907). "Rise and Fall of N Rays; The History of a Belief", *L' Année Psychologique,* Vol. 13, pp. 143-169.

Rostand, J. (1958). *Science Fausse et Fausses Sciences,* Gallimard, Paris.

Seabrook, W. (1941). *Doctor Wood,* Harcourt, Brace and Co., New York.

Vallery-Radot, R. (1923). *The Life of Pasteur,* Doubleday, Page and Co., New York.

Widener, A. (1979). *Gustave Le Bon,* Liberty Press, Indianapolis, Ind.

Wood, R. W. (1904). "The n-Rays", *Nature,* Volume 70, pp. 530-531.

Wood, R. W. (1912). "Resonance Spectra of Iodine by Multiplex Excitation", *Philosophical Magazine,* Volume 24, pp. 673-693.

Wood, R. W. (1917). *How to Tell the Birds from the Flowers and Other Woodcuts.* Dodd, Mead and Co., New York. Reprinted in 1959 by Dover Publications, Inc., New York.

IV
Grand Illusions: Russian Water

Water is unique. It is the one chemical substance with which everyone is familiar. However vague our scientific conceptions, each of
us recognizes by direct experience the vital role played by water in
our very existence. It is in us and around us. It comes as no surprise,
therefore, to learn that Thales, the man Aristotle called the founder
of Western philosophy of science, asserted that water is the source
of all things: υδωρ παντα (water is all).

Variations of this view prevailed for millennia. It has always been
apparent to anyone watching an open container of water that the
liquid disappears slowly; that the water is converted into air seems
obvious. It has also been apparent for millennia that living organisms
imbibe water and increase in size and weight; clearly water can be
converted into solids.

In fact, a "definitive" experiment that proved that water could
be converted into solids was carried out in recent times, the seven-
teenth century, by a distinguished savant, Joan Baptista van Helmont.
His procedure was a quantitative, precise one, very modern in its
conception and execution.

I took an earthen vessel, in which I put 200 pounds of earth that had been
dried in a furnace, which I moistened with rainwater, and I implanted therein
the trunk or stem of a willow tree, weighing five pounds. And at length,
five years being finished, the tree sprung from thence did weigh 169 pounds
and about three ounces. When there was need, I always moistened the earthen
vessel with rainwater or distilled water, and the vessel was large and implanted
in the earth. Lest the dust that flew about should be co-mingled with the
earth, I covered the lip or mouth of the vessel with an iron plate covered
with tin and easily passable with many holes. I computed not the weight
of the leaves that fell off in the four autumns. At length, I again dried the

earth of the vessel, and there was found the same 200 pounds, wanting about two ounces. *Therefore 164 pounds of wood, bark and roots arose out of water only.*

Obviously he had transmuted water into earth-like matter. van Helmont also recognized that water could be converted into air. Thus he rejected the "heathen" theory of four elements (earth, air, water, fire); to him water was the fundamental element. *Wasser ist alles* (water is all).

Van Helmont's experiment was later repeated by Boyle who confirmed the former's observation. Boyle then went a step further, simplifying the design of the experiment: he grew some small plants in water alone. Again finding the plants had gained in weight, and size, he concluded that water was transmuted into the plant and into the various substances that he was able to isolate from the plant. Furthermore, Boyle argued:

> the plants my trials afforded me, as they were like in so many other respects to the rest of the plants of the same denomination; so they would, in case I had reduced them to putrefaction, have likewise produced worms or other insects as well as the resembling vegetables are wont to do; *so that water may, by various seminal principles, be successively transmuted into both plants and animals.*

In other words, *water is all* (water is all).

In modern science, water has lost its position as a fundamental element. However, its importance and centrality in science has increased rather than diminished. It plays a preeminent role in physics, chemistry, geology, geophysics and the life sciences. In physics it provides the reference standards for fundamental quantities. In chemistry it is the ubiquitous solvent for innumerable reactions. In geology and geophysics it is the most important terrestrial agent by which the surface of the earth is modified. In biology it is the matrix in which all living organisms maintain themselves and reproduce.

Under these circumstances it is no surprise that enormous efforts have been devoted, in the modern era of the past two centuries of chemistry, to explore exhaustively the properties and behavior of water. Thus more is known about water than about any single chemical entity. Nevertheless, as our knowledge increases, our understanding does not progress proportionately, because we ask increasingly more sophisticated and detailed questions.

Water constitutes about 0.2 percent of the volume of the earth, primarily in the oceans, but also in the inland lakes and rivers and, less obviously, absorbed or submerged by the soil. At the temperatures on earth, water is a very inert chemical substance, as it would have to be to fulfill its role in geological and life processes. On the other hand it has a collection of characteristics that make it uniquely fit as a central agent of geological and biological transformations. As L. J. Henderson said about seventy years ago:

> In truth Darwinian fitness is a perfectly reciprocal relationship. In the world of modern science a fit organism inhabits a fit environment....Water, of its very nature, as it occurs automatically in the process of cosmic evolution, is fit, with a fitness no less marvelous and varied than that fitness of the organism which has been won by the process of adaptation in the course of organic evolution.

Let us review some of the exceptional characteristics of water. The most familiar are some of the thermal properties. Water has an exceptionally large capacity to absorb heat; in technical terms, it has a high specific heat (1.0, compared to 0.1 for iron, 0.2 for glass, 0.3 for sugar). This is of immense significance in moderating the climate on earth. The effects of this heat capacity are strikingly illustrated in a comparison of marine and continental temperatures (Figure 1).

The large heat capacity heat is also of crucial significance for human physiology. The energy produced by an average adult human being at rest is about 2,500,000 calories. If the heat-absorbing capacity of the body were that of most chemical substances, metabolic energy would raise the temperature of an adult by over 100° Centigrade. Because the body is constituted mostly of water, the potential rise in temperature is only 30°. The load on our cooling mechanisms is markedly reduced. Since increased temperature has a striking effect on chemical reactions (a rough rule of thumb is that the velocity of reaction doubles for every 10° increase in temperature), large fluctuations would have a disastrous impact on a living organism. For this reason very sensitive homeostatic mechanisms have evolved in higher organisms, and environmental temperature modulation is important for more primitive organisms. Alternatively, one can use creationist rationalizations and argue that these characteristics are evidence of Design in Nature. In either interpretation, the large heat capacity of water is the crucial feature.

Two other thermal properties of water, the heat of melting (80 calories per gram) and the heat of evaporating or boiling (536 calories per gram) have exceptionally high values, in fact substantially larger than those of practically any other known material. These charac-

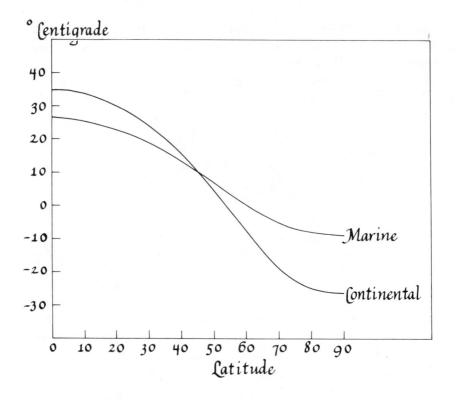

Figure 1. Variation of average temperature of land and of oceans at different latitudes.

teristics lead to striking consequences. If 30 calories of heat are extracted from one gram of water at 30° Centrigrade, the temperature of the liquid will drop 30°, that is to 0° Centigrade. However, if an additional 30 calories of heat are removed from the water, its temperature will not drop at all, but will stay fixed at 0°. Instead, part of the liquid will freeze. In fact 80 calories must be extracted from the liquid to convert it totally to ice, and thereafter the temperature can be reduced below 0° Centigrade. Thus water serves as an enormous reservoir of heat to protect life on earth. Practically, the cooling of oceans cannot progress below about 0° (actually about -1° Centigrade because of the salt therein) and hence, barring a super-cold catastrophe that would extinguish all life on earth, the aqueous ocean can be a secure haven for living organisms.

Complementary behavior occurs as water is warmed. If a body of water is open and exposed to air, the absorption of energy, from the sun, for example, will raise its temperature by only a relatively small amount; the largest portion of the heat absorbed from the sun is expended in evaporation of the water. Each year, at the equator, about 10^{15} calories of heat from the sun absorbed per square kilometer are used to evaporate a quantity of water equivalent to over two meters in depth. The heat absorbed in this evaporation of water in the tropics is released later when the water-saturated air has been carried to a cooler place and turns into rain. No other liquid could be as effective in transporting heat, by means of its vapor, from a warm locality to a colder one.

Evaporation of water also plays a dominant role in the physiological regulation of body temperature. As the ambient temperature increases, the heat loss from the skin and lungs of a human being approaches 80 percent of the total heat "excreted." The elimination of this excretory product is as essential for health as is the excretion of liquids and solids.

The freezing temperature of water, 0° Centigrade, and the normal boiling temperature, 100° Centigrade, are also surprisingly high for molecules of that molecular weight (18 for H_2O). Thus neon (Ne, molecular weight 20) boils near $-250°$, methane (CH_4, molecular weight 16) near $-160°$ and ammonia (NH_3, molecular weight 17) at $-33°$. Obviously H_2O boiling at 100° is exceptional among light molecules. At the very cold temperatures at which methane or ammonia are liquids, the velocities of most chemical processes would be extremely slow.

In contrast to practically all other liquids, water manifests the peculiar property of *expanding* in volume when it is cooled below 4° Centigrade. Above 4°, it expands as it is warmed. Thus the volume of the liquid is at a minimum at 4°.

Related to this extraordinary volume behavior of the liquid is the expansion in volume when water is frozen. It has been recognized for several centuries that the buoyancy of ice and the expansion of water below 4° are responsible for the liquid state of most large bodies of water in frigid climates. If the coldest water were the densest, it would sink to the bottom of a lake, and if the ice were still more dense, it would settle to the bottom. Any warm water formed in summer would float near the surface and not perturb the ice at the bottom. Progressively over the years more ice would form and persist in the lower region of a lake even through summers, so that ultimately the lake would become a massive block of ice. All life therein would perish. About 200 years ago, Prout, being aware of these characteristics of water, concluded as follows:

> The above anomalous properties of the expansion of water and its consequences have always struck us as presenting the most remarkable instances of design in the whole order of nature — an instance of something done expressly, and almost (could we indeed conceive such a thing of the Deity) [as] a second thought, to accomplish a particular object.

Turning from thermal to other properties of water, we recognize that this liquid is a remarkably versatile solvent for an enormous range of inorganic, organic and biological substances, ranging in molecular size from small molecules such as ammonia (NH_3, molecular weight 17) and methane (CH_4, molecular weight 16) to giant macromolecules such as hemoglobin (molecular weight 65,000) or synthetic polyvinylpyrrolidone (molecular weight 1,000,000). Large quantities of these solutes dissolve in water. It is the only solvent that could serve to carry the enormous variety of solutes in blood, or in urine.

If one also includes trace amounts, then even sand (silica) or rust (iron oxide) can be dissolved by water. This latter property, as well as water's mechanical action, are the basis of its monumental role in geology. No other liquid would have been as effective in geological processes.

Water also possesses remarkable electrochemical properties. It has a surprisingly high dielectric constant (80). Comparable small-molecule liquids, e.g., ammonia or alcohol, show much lower values (16 and 22, respectively). A high dielectric constant diminishes the electrostatic attractive forces between oppositely-charged particles. This feature as well as its ability to solvate charged species make water the most powerful solvent for ionization of salts. The electrical characteristics of ionic solutions are the basis of phenomena as diverse as storage batteries and nerve impulses.

Water is also remarkably fit as an environment when viewed at the molecular level as well as at the organism and planetary levels. Most astonishing among its properties is the velocity at which hydrogen ions, H^+, and hydroxide ions, OH^-, can travel in water and particularly in an ice matrix. Contrary to wide experience with other liquids, an H^+ ion moves *faster in solid ice than in liquid water.* The rate at which H^+ and OH^- can move toward each other in liquid water (10^{-11} sec for collision), or ice (10^{-12} sec for collision), is *faster* than that at which any two other solute species can diffuse toward each other by general thermal movements. Also surprising among the eccentricities of water is the fact that the fluidity of the liquid *increases* with an increase of external pressure. One would expect the molecules to be packed closer together at high pressures, and to have a more difficult time slipping past each other.

For almost a century it has been a challenge to understand the unusual properties of water in terms of its molecular structure. To say that water is H_2O tells us little beyond its chemical composition and the stoichiometry of its chemical reactions. This was aptly expressed by a famous British examination question: " 'To the chemist water is H_2O, just H_2O' (Oxford Sermon). Discuss." What we need to know is how H_2O molecules are arranged in the solid and liquid phases, as well as the distances between constituent atoms in the gas phase.

Although spectroscopy has played an important role in establishing interatomic distances and angles in an isolated H_2O molecule, it is X-ray diffraction that provided the crucial data for defining the spatial arrangements of water molecules in the solid and liquid states. The single most influential paper in this area was written in 1933 by J. D. Bernal, a brilliant, pioneering X-ray crystallographer and molecular structurist, and R. H. Fowler, a renowned theoretical physicist. As fate would have it, Bernal reappears in our story of polywater, forty years later, in two different cloaks.

John Desmond Bernal (Figure 2) was one of the many gifted sons of Ireland who became outstanding in British arts and science. Brilliant, erudite, articulate and immensely gifted in many areas, he played a pioneering role in elucidating the molecular structure of molecules from water to viruses. Two of his students, Dorothy Hodgkin and Max Perutz, received Nobel Prizes for determining the molecular structures of large biological molecules, but it was Bernal who first showed, thirty years earlier (in the 1930s), that this was a feasible goal. He was also an avowed Marxist and staunch friend of the Soviet Union, even through the scientific and political turmoil there in the postwar era. He was awarded the Soviet Lenin Peace Prize in 1953. He was an honorary professor at the University of Moscow, and a foreign member of the National Academy of Sciences of the USSR, of Poland, of Hungary, of Rumania, of Bulgaria, and of Czechoslovakia. Among his many books is *Marx and Science,* which he wrote in 1952. Despite his far-left politics, he was asked to join the British military operations analysis group during World War II. Among other successes, he played a crucial

Figure 2. John Desmond Bernal (1901–1971) examining a model for liquid structure, sometime in the early 1960s. (This photograph and permission to use it kindly provided by Dr. J. L. Finney of the Department of Crystallography, Birkbeck College, University of London, where Bernal spent most of his scientific life.)

role with Field Marshal Montgomery, particularly in deciding the disposition of fire power and armor against Rommel in the turn-around at El Alamein and in the final defeat of the Nazis in North Africa.

The Bernal-Fowler paper set down the basic principles of the structure of water. The central point is that each water molecule, H-O-H, tends to be surrounded by and bonded to four other water molecules placed at the vertices of a tetrahedron (Figure 3). Between each pair of oxygen atoms (filled circles in Figure 3) is a hydrogen

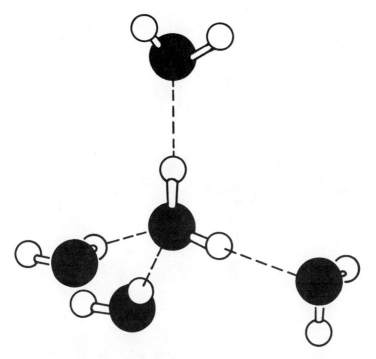

Figure 3. Tetrahedral arrangement of the four water molecules around a central water molecule. The larger dark spheres represent the oxygen atoms and the smaller open spheres the hydrogen atoms in each H_2O.

atom. In normal ice the O-H$\cdots$O distance is 2.76 Å, in liquid water somewhat longer (2.88 Å at room temperature). Looking at a larger collection of water molecules in continuing tetrahedral arrangement, one finds hexagonal cages. In Figure 4, there is an O atom at every vertex, and every line between vertices has an O-H$\cdots$O "hydrogen bond" between oxygens. In normal ice the boat and chair hexagons of Figure 4 tend to be nearly perfect and fixed in geometry. In liquid water, defects, distortions and deformations exist, which disrupt the highly regular array of the solid state. The structure in any local region in the liquid also fluctuates with time.

Figure 4 illustrates the disposition of H_2O molecules in *one* form of solid ice, the one familiar to all of us, designated ice I. There are eleven solid ices (all known before Kurt Vonnegut invented his special polymorph of ice-nine in *Cat's Cradle*). These ices are not familiar to most people because they exist only at very high pressures. For example, ice VII can be obtained at pressures near 300,000 pounds per square inch, but in this extraordinary environment the crystal is stable even above 50° Centigrade.

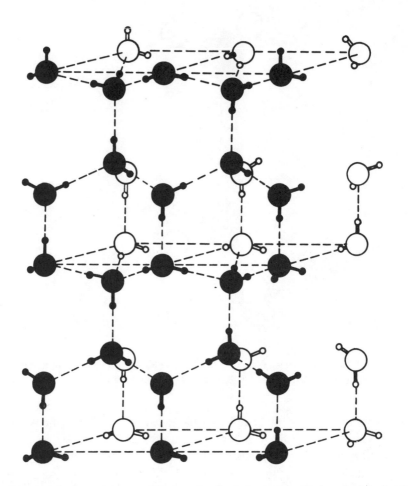

Figure 4. Hexagonal arrays of H_2O molecules in ice, originally described in detail by J. D. Bernal and R. H. Fowler in 1933. With some distortions, defects, and bending, the model also can represent the arrangement of H_2O molecules in liquid water.

The high pressure ices are much more dense than ice I. For the latter the density is 0.92 grams per cubic centimeter, for ice II it is 1.18, for ice VII it is 1.56. In these, the arrangement of H_2O molecules must be more densely packed to put more mass in a given volume. Figure 4, the model for ice I, has much empty space in it. X-ray crystallography of the high pressure ices has shown that the H_2O molecules are arranged in intertwining networks that fill in much of the open space of the tridymite diamond-like lattice (Figure 4). For example, each H_2O molecule in ice VII is surrounded by eight others, at equal distances.

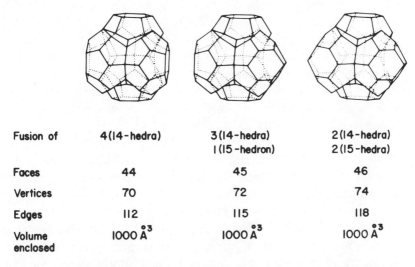

HYDRATE POLYHEDRA

	Dodecahedron	Tetrakai decahedron	Pentakai decahedron	Hexakai decahedron
Faces	12	14	15	16
Vertices	20	24	26	28
Edges	30	36	39	42
Volume enclosed	160 $\overset{\circ}{A}{}^3$	230 $\overset{\circ}{A}{}^3$	260 $\overset{\circ}{A}{}^3$	290 $\overset{\circ}{A}{}^3$

MULTIPLE FUSED POLYHEDRA

Fusion of	4 (14-hedra)	3 (14-hedra) 1 (15-hedron)	2 (14-hedra) 2 (15-hedra)
Faces	44	45	46
Vertices	70	72	74
Edges	112	115	118
Volume enclosed	1000 $\overset{\circ}{A}{}^3$	1000 $\overset{\circ}{A}{}^3$	1000 $\overset{\circ}{A}{}^3$

Figure 5. A selection of types of water polyhedra with enclosures of varying internal volume, from 160 Å³ to 1000 Å³. Other known types can enclose even larger volumes. At each vertex of a polyhedron there is an O atom. Between adjacent vertices, that is, along each edge, there is a hydrogen atom in an O-H···O bond. These clathrate hydrates form enclosures for guest molecules of a wide range of structure, for example, argon (Ar), chlorine (Cl₂), chloroform (CHCl₃), benzene (C₆H₆), tetraisoamylammonium fluoride [(i-C₅H₁₁)₄NF], etc. Such polyhedral hydrates exist as beautiful crystals, often near 90 percent water in composition and yet not melting until the temperature is substantially above 0° Centigrade, the melting temperature of common pure ice I.

So far we have mentioned only ices made from pure water. It is also possible to obtain even more novel "ices" from water to which a small amount of "impurity" has been added. These are called polyhedral "clathrate hydrates." The first one, chlorine hydrate, was discovered by Humphrey Davy and by Michael Faraday at the beginning of the nineteenth century. Subsequently, the inert gases, argon, krypton, and xenon were found to form crystalline stoichiometric hydrates, and by now about a hundred polyhedral hydrates have been isolated. In composition they are overwhelmingly water. X-ray crystallogrpahy has shown that the "impurity" or guest molecule is enclosed in a cage of water molecules. A few of the many different known cages are shown in Figure 5. It is apparent that many different polyhedra can be formed, with pentagonal (and even quadrilateral) as well as hexagonal faces, and with cavities for the enclosure of a wide range of sizes of guest molecules.

Obviously water is a remarkably versatile material, at the molecular as well as the macroscopic level. Against this background we will examine the impact of the discovery of polywater, and the different human responses to the announcement thereof.

Although descriptions had appeared earlier in the Soviet literature, anomalous water began to receive attention in Europe and the United States only after the publication in 1966 of a paper by Derjaguin in the *Discussions of the Faraday Society*. This paper carried the unassuming title "Effect of Lyophile Surfaces on the Properties of Boundary Liquid Films." This title gave no hint of a revolutionary discovery; it was essentially reiterative of the kinds of research on surfaces for which the author was well known. However, the paper was actually presented at a Faraday Society conference of colloid and surface physical chemists in Nottingham, and at that meeting the seed of polywater was planted and started to grow.

Boris Vladimirovich Derjaguin (Figure 6) is a Russian physical chemist of distinction, a (corresponding) member of the USSR Academy of Sciences and head of the Laboratory of Surface Phenomena at the Institute of Physical Chemistry of the Academy, in Moscow. His research has focused on the physics and chemistry of surfaces, particularly on the properties of thin layers of fluids (which are of practical importance in the field of lubrication). He was well known for his work on colloids, as well as thin liquids, and was a very appropriate scientist to try to entice from behind the Iron Curtain to an international conference of the type sponsored periodically by the Faraday Society.

Figure 6. Boris Vladimirovich Derjaguin (b. 1902) leader of the Soviet research group that first directed attention to a "new" form of liquid water with extraordinary characteristics. (Reprinted from Chemical & Engineering News, July 16, 1973, vol. 51, p. 13.)

In his Nottingham presentation, Derjaguin described some interesting experiments on the condensation of water into capillary tubes of tiny diameter (a few millionths of a meter) when the vapor pressure of water in the gas phase above the capillary is *below* saturation. He credited the initial observation to a lone Soviet scientist, N. N. Fedyakin (originally from the Technical Institute of Kostroma, 200 miles from Moscow) whose name appears as a coauthor on some of the water papers from Derjaguin's laboratory, but who was an elusive figure both before and after this incident. Then Derjaguin described much more elegant equipment for studying the phenomenon, and much more extensive experimentation confirming and extending Fedyakin's observation.

That water in the gas phase at a pressure below the saturation vapor pressure of the liquid should condense into the liquid phase is astonishing. Taken at face value this observation violates the laws of thermodynamics; if true, it means one could construct a perpetual motion machine. Such a conclusion is essentially unthinkable, even in the modern world of relativity and quantum mechanics. In fact Einstein himself said:

> [*Thermodynamics*] is the only physical theory of universal content concerning which I am convinced that, within the framework of the applicability of its basic concepts, it *will never be overthrown.*

The alternative conclusion is that the "dew point," the saturation vapor pressure of the water in the capillary tube, is *below* that of ordinary liquid water. In that case, either the glass capillary is exerting some special effect on the water, or the water in the capillary is different. A century ago, Lord Kelvin demonstrated on thermodynamic grounds that a very tiny-bore capillary would affect the vapor pressure of a pure liquid within it and derived a mathematical equation to relate the vapor pressure to the capillary diameter. In principle, such effects should appear in tubes of less than one-millionth of a meter in diameter. Derjaguin demonstrated convincingly that the Kelvin effect could not be the basis of his observations by comparing a series of capillaries of different bores and obtaining the results he tabulated as follows:

Capillary radius $= r$	$\dfrac{\text{Vapor pressure capillary water}}{\text{Vapor pressure normal water}} = \dfrac{p}{p_s}$
3	0.93
5	0.93
12	0.93
18	0.93
22	0.93

As Derjaguin said in his article: "The absence of any perceptible dependence of p/p_s on r indicates that in these experiments there were bulk modifications of the...liquid" and that the glass (or quartz) container played only a peripheral role, facilitation of the formation of this new phase.

Having convinced himself of the existence of a new form of liquid water, Derjaguin in his Nottingham paper proceeded to describe some of its unusual properties. It had a viscosity fifteen times that of normal water, a thermal expansion (in the 20–40°Centigrade range), one and one half times greater, a boiling point somewhere above 150°Centigrade, unusual freezing behavior in the range of −15° to −30°Centigrade. Therefore, he concluded, "the usual state of water...is thermodynamically metastable. Therefore...it would be convenient to call usual water metawater, and the anomalous columns — *orthowater.*"

Alternative names used by Derjaguin were *anomalous water* and *water II.* Ultimately the new phase was most commonly called *polywater,* a term invented by an American, Ellis Lippincott, to include structural presumptions.

Somewhat parenthetically, I should mention one feature that was puzzling even at the time, but seems never to have been pursued. The table that presented the lowered vapor pressures of anomalous water also contained comparable information for three other liquids: acetone, methyl alcohol, and acetic acid. In every case, vapor phase transfer of these liquids from a reservoir to small quartz capillaries produced a phase with vapor pressure *lower* than that of the normal liquid in the reservoir. Following the same argument as used for water, one must conclude that Derjaguin had created ortho-acetone, ortho-methanol and ortho-acetic acid, respectively. Nevertheless, at no time was any excitement generated and nowhere do we find glowing accounts of polyacetone, polymethanol or polyacetic acid, either in the scientific literature or in newspapers.

In papers other than the Nottingham one, Derjaguin and his colleagues described more unusual properties of polywater. It had a density of 1.4, a refractive index of 1.46, compared to 1.0 and 1.33, respectively, for normal water. Its viscosity was comparable to that of motor oil. It did *not* show a minimum in volume at 4°Centigrade, but continued to decrease in volume with decreasing temperature down to −12°Centigrade. Its electrical conductance was higher than that of ordinary water. Further stability tests indicated the new phase could be heated to over 400°Centigrade without losing its anomalous properties.

Scientists who heard of Derjaguin's new discoveries and had sufficient motive to make a judgment were buffeted by many influences. First there is the basic dilemma: is the claim valid or suspect. We are constantly being told that we are refractory toward new ideas. At the very moment that this paragraph was being written, the greatest chemist of this century, Linus Pauling, was complaining to the journal *Science* that, his pet peeve, the medical establishment, "doesn't know how to recognize new ideas that are worthwhile." On the other hand, for millennia we have all been cautioned by the biblical injunction: "beware of false prophets."

Which assessment was appropriate at the time of the disclosure of Derjaguin's claims? There is no litmus test for instantaneous discrimination. In general one must wait for the test of time.

A number of individuals formed tentative opinions on the basis of various non-objective criteria that we all use in reaching everyday decisions. One of these is, what is the reputation of the claimant? Derjaguin's was not unblemished. His past experiments were generally very difficult to execute and hence not attractive to repetition in other laboratories. His views on long-range forces between surfaces in juxtaposition and on the properties of fluids in pores of granular material were viewed in many quarters as more romantic than logically compelling. Some of his claims seemed contrary to established physicochemical principles. One of Derjaguin's counterparts in the United States was Curtis Singleterry, in charge of similar researches at the Naval Research Laboratory. This highly respected individual had been skeptical of Derjaguin's previous interpretations as well as of his disclosure of anomalous water, although he considered the Russian's experimental designs in colloid and surface chemistry imaginative. Those who knew Curtis Singleterry, the epitome of a rational scientist and self-secure, modest human being, could not but be impressed by his critical views, which, typical of the man, he did not express in print.

Others felt uneasy with the description of the discovery. The grand old man of the physical chemistry of liquids, Joel Hildebrand (whose one hundredth birthday was celebrated in 1982 on the Berkeley campus), expressed his doubts in a short piece of somewhat "creative" writing:

> Proponents of polywater in the pages of *Science* and elsewhere may be interested to learn why some of us find their product hard to swallow. One reason is that we are skeptical about the contents of a container whose label bears a novel name but no clear description of the contents. Another is that we are suspicious of the nature of an allegedly pure liquid that can be prepared only by certain persons in such a strange way. We choke on the explanation that glass can catalyze water into a more stable phase. Water and silica have been in intimate contact in vast amounts for millions of years; it is hard to understand why any ordinary water should be left.

On the other side of the fence we found many who deeply hoped that Derjaguin was right. During a visit of Derjaguin to Bernal's laboratory at Birkbeck College (University of London), the host is recorded as saying, with regard to polywater,

> In my opinion this is the most important physical-chemical discovery of the century.

Derjaguin responded:

> I am glad to hear you say this: I would like to ask you something. Would it be possible for you to write something later about your opinion on the significance of this work.... It would be very important for me to get such an estimate.

Bernal said, "I will be glad to do this." Clearly he very much wanted a Soviet scientist to receive recognition for a great fundamental discovery.

Other emotional responses were also instigated by political views. Anti-American feelings in some quarters in Great Britain led to hopes that the Russians would outdo the Americans in some area of science. In such quarters, the skepticism of many American scientists was attributed to envy and unhappiness that a Russian had made a potential Nobel Prize–winning discovery. Anti-American hostility often expressed itself in refusal to disclose anything to scientific visitors from the London office of the U.S. Office of Naval Research, although if the same visitor, generally an academic scientist on a one-year sabbatical in London, went there as a visiting professor he would be received civilly.

Scientific controversies, however, are not resolved by a Gallup poll or by any counting up of individuals pro or con. History shows too many instances in which majority opinion, even of experts, was wrong. There is plenty of precedent for discounting a preponderant view.

About a century ago, young Svante Arrhenius encountered massive bitter opposition when he proposed that salt dissolved in water becomes separated into two oppositely-charged entities, Na^+ and Cl^-. And rightly so. After all, as any high-school student knows, oppositely-charged entities attract each other, as specified by Coulomb's law, with a force proportional to the magnitude of the charges q_+ and q_-, respectively, and inversely proportional to the square of the distance, r, between them:

$$\text{Force} = \frac{(q_+)\,(q_-)}{r^2}$$

True this equation applies strictly in a vacuum, but in a solvent the change would be merely the insertion of the dielectric constant, D, in the denominator, not a reversal in the attractive character of the force. For water D is near 80, so the attractive force would be weakened, but not abolished. Why should positive Na^+ and negative Cl^- stay apart as separate species? Furthermore, it was hard to believe that in water sodium could exist separately from chlorine, for after all separate pure sodium reacts violently when placed in water. Also pure chlorine separated from sodium is toxic, whereas saltwater is not.

Arrhenius presented his ideas in his doctoral thesis — which was almost rejected — in 1883. After much discussion, his examiners at the University of Uppsala grudgingly agreed to give him a *fourth class* doctor's degree, perhaps because his earlier performance at the university was a distinguished one. This pass was so low (equivalent to a D grade in the United States) that it did not qualify him to become a beginning faculty member, a docent. The opinion of the Uppsala faculty was subsequently confirmed by then outstanding British chemists, such as S. U. Pickering, J. H. Gladstone and G. F. Fitzgerald, who criticized Arrhenius's ideas vigorously at an 1890 meeting of the British Association for the Advancement of Science on "Theories of Solution." Fortunately, Wilhelm Ostwald, then at the University of Riga and one of the giants of physical chemistry, was so impressed by Arrhenius that he offered the young man an appointment as docent at Riga. It seems to be a universal law in academic life that if some other university wants you, then you must be good. So Arrhenius was given a junior appointment at Uppsala. Soon Van't Hoff also supported Arrhenius's views and the two published jointly in 1887 in the first issue of the *Zeitschrift für physikalische Chemie.* The Swedes still were not very much impressed and it was only with reluctance, and in the absence of an alternative candidate, that in 1895 Arrhenius was given a professorship in Stockholm.

Twenty years after receiving his doctorate degree, Arrhenius was awarded the Nobel Prize in chemistry. He was third in line, after J. H. Van't Hoff and Emil Fischer. Except for a few diehards who maintained strong and vocal attacks on ionization theory, the scientific world forgot its opposition to Arrhenius and instead continued to shower honors upon him.

Nowadays it is difficult to appreciate how ferocious and malignant was the antagonism to relativity theory. In some circles the antipathy expressed majority opinion, and was vicious in its open hatred. Philipp Lenard, the distinguished German physicist, wrote:

> [Relativity is] a Jewish fraud, which one could have suspected from the first with more racial knowledge than was then disseminated, since its originator, Einstein was a Jew....The most important example of the dangerous influence of Jewish circles on the study of nature has been provided by *Herr Einstein with his mathematically botched-up theories* consisting of some ancient ideas and a few arbitrary additions. *This theory now gradually falls to pieces.*

Lenard was not alone in this view. In 1931, even before the Nazis took over the German government, a book appeared entitled *One Hundred Authors Against Einstein* which expressed similar criticisms (as one wit pointed out one hundred times zero is still zero). While raging against Einstein as a fraud, Lenard simultaneously claimed that Einstein's ideas had been anticipated by a certified Aryan physicist, Friedrich Hasenöhrl (a theoretician at the University of Vienna, killed at the front in World War I). In particular, Lenard found a percursor to the famous equation $E = mc^2$ in the work of Hasenöhrl. Consequently Lenard placed Hasenöhrl in a class with Galileo and Newton. During the Nazi period German theoretical physicists managed to use relativity theory by referring euphemistically to "Hasenöhrl's principle."

Since plebiscites would be unconvincing, the opposing polywater camps began to turn from pure talk to work. For those who were favorably inclined, it was essential to obtain experimental and theoretical evidence that polywater existed. For those who were skeptical the object was to demonstrate that the liquid prepared by Derjaguin's procedures was not a pure phase but a complicated mixture.

A major boost to polywater proponents was provided by Ellis Lippincott and his co-workers when they succeeded in obtaining an infrared spectrum of the tiny amount of liquid collected in a small capillary. What was striking about the published spectrum was the absence of an absorption peak near 3400 wave numbers that is so characteristic of normal liquid water. Instead a strong peak was found much deeper in the infrared region, near 1600 wave numbers. Very properly, Lippincott interpreted this drastic shift as diagnostic of the formation of very strong $O-H \cdots O$ bonds between water molecules. In fact by analogy with other very strong hydrogen bonds, $F-H \cdots F$ in alkali metal fluoride salts, Lippincott proposed that the oxygen to oxygen distance was 2.3 Å in polywater, in contrast to 2.8 Å in the normal liquid.

Shortly thereafter the presence of strong hydrogen bonds in poly-water was bolstered by observations in nuclear magnetic resonance experiments, in which again a peak was found that was very much displaced from that characteristic of normal water, in a direction consistent with stronger hydrogen bonding.

Such experiments, as well as a preliminary (later unsubstantiated) report of a distinctive pattern in X-ray diffraction generated much enthusiasm among theoretical chemists. A variety of quantum mechanical calculations were made, all leading to structures with stronger and shorter O-H$\cdots$O bonds. One of the more excitable and optimistic theoreticians was so elated with his results that he was prompted to say (to his later regret in his more reflective, critical moments): "We have presented arguments, supported by quantum mechanical calculations, which we believe *establish* [polywater's] existence and characterize its properties." In general the quantum theoreticians favored a structure for polywater that was similar in atomic arrangement to the planar hexagons found for carbon in graphite. For normal water the puckered ring structure is similar to that for carbon in diamond.

In parallel with progress among the enthusiasts, experimental work casting doubt on the purported nature of polywater began to appear increasingly.

From the very beginning there were questions about the purity of the new material. These lurked even in Derjaguin's mind. In the Nottingham paper he explicitly recognized the possibility of impurities from glass coming into the aqueous phase in the capillary but he claimed this factor could be dismissed by the observation of similar behavior in quartz capillaries. This claim was never convincing, however, as Hildebrand indicated:

> There is another and, I think, much more plausible role for the necessary glass. Water and silica interact in wonderful variety, as may be read in a fascinating book by Ralph K. Iler, *The Colloid Chemistry of Silica and Silicates* (Cornell University Press, Ithaca, N.Y., 1955). It is easy to see why a spectroscopist might be excited by the term "polywater" to try to design new ways for water to polymerize which nature had overlooked, but I think that a chemist who feels curious about what is in those glass capillaries would have more success if he assumes that he is dealing with a system of two components.

In time even Bernal moved into Hildebrand's camp, for Bernal's perceptive mind took control of his heart and he wrote: "One of the greatest difficulties in even accepting the existence of a more stable phase [of water] is its apparent absence in nature."

Despite Herculean efforts, nobody succeeded in preparing macroscopic quantities of polywater. There was fairly wide success in obtaining quantities of the order of a millionth of a gram inside the tiny capillaries and in extruding microscopic droplets therefrom. Bernal's group tried valiantly to obtain enough polywater for X-ray diffraction studies. Unilever in Great Britain devoted much thought and effort to scaling up the production of water. In the United States, the Department of Defense, through the Advanced Projects Research Agency, sponsored at least one large project for the production of polywater. Although it is easy to smile sardonically now at the presumed gullibility of a defense agency, the fact is that no intelligent administrator in a responsible position could have taken the risk of dismissing the Russian discovery out of hand. After all a significant number of reputable American scientists were convinced that Soviet scientists had a new form of water more stable then the common phase; what if. . .? Surely a few hundred thousand dollars for a crash investigation was a prudent expenditure.

None of these, or other, attempts to obtain macroscopic quantities of polywater was successful. Even in the Soviet Union, all such attempts failed as is evident from the following, slightly humorous, exchange (reported in a Russian Journal) between Derjaguin and A. N. Frumkin, the doyen of Russian surface science:

Frumkin: How much modified water has been obtained in all?
Derjaguin: About enough for fifteen articles.

Critical investigators had to work, therefore, with microgram quantities, and they adapted a number of analytical techniques for their purposes. Analysis in the United States of polywater samples with electron microprobes showed the presence of sodium and boron. Analysis by electron spectroscopy disclosed the presence of sodium, potassium, sulfate, chloride, borate and carbonate.

Derjaguin's rejoinder, when he was presented with these observations, was an undeviating one, such as that in a 1971 paper delivered at a symposium of the American Chemical Society: "Unfortunately, many authors have obtained conflicting data for materials prepared without careful experimental considerations." In other words other people's samples contained impurities because of sloppy techniques, but his material was pure.

Actually analyses on Derjaguin's samples per se had been carried out in Moscow, by V. L. Tal'rose of the Institute of Chemical Physics of the USSR Academy of Sciences, as early as 1968. Using mass spectrometry and sub-microgram amounts of polywater, Tal'rose had found large quantities of impurities in the liquid, particularly organic materials such as phospholipids and lipids. In a complementary direction, the mass spectra showed only (charged) monomeric H_2O molecules, no polymeric $(H_2O)_n$ species. Admittedly these observations were never published in a formal paper; they were publicly disclosed, yet not alluded to by Derjaguin.

Other damaging evidence also began to appear. Since deuterium (D) atoms are twice as heavy as H atoms, one would expect the atomic vibrations of O-D$\cdots$O in D_2O, heavy water, to be different from those of O-H$\cdots$O in H_2O water. If these vibrations were different, the infrared spectrum of polywater prepared from D_2O should be different from that prepared from H_2O. R. E. Davis, from Purdue University, reported, however, that these infrared spectra had the same features, a result incompatible with an assignment to hydrogen vibrations. He was inspired to produce the following piece of creative writing:

> Thus we must conclude that all polywater is polycrap and that the American scientists have been wasting their time studying this subject unless, of course, it can be defined as a topic of water pollution and waste disposal.

At about the same time, S. W. Rabideau and A. E. Florin from the Los Alamos Laboratory in New Mexico repeated the nuclear magnetic resonance experiments published by others and very sensibly carried out a necessary set of control scans — with an *empty* capillary. They reported,

> Although an apparent broad absorption signal was observed [for polywater in a capillary] approximately 300 Hz downfield from the ordinary proton resonance, a *corresponding number of scans* at the same radio-frequency levels *with an empty capillary gave a hump in this same region.*

Such control scans had been omitted previously. Clearly the anomalous signal attributed to polywater has its origin in material in the glass container. In fact it had been shown by others previously that protons adsorbed on silica gel exhibit such a signal.

Simultaneously several investigators reported that mixtures of silica (the major constituent of glass or quartz) and water showed properties very similar to those attributed to polywater, as in fact

Hildebrand had expected. It became increasingly evident that the unusual properties of polywater were due to the impurities in it.

Suddenly and astonishingly, Derjaguin himself conceded that polywater was really normal water with impurities leached from the quartz in which it was prepared. He himself had carried out analyses of the contents of the capillaries and found dissolved silica. The evidence had become overwhelming. In 1973, in a short note to *Nature,* he wrote:

> Consequently, the anomalous properties of condensates may be explained, not by the formation of a new modification of water, as was previously supposed, but by the peculiar features of a reaction taking place between the vapour and solid surfaces in the process of condensation.

One must admire a man who had the intellectual honesty and emotional courage to triumph over a deep emotional and intellectual commitment, who was willing to face the public embarrassment, in the Soviet Union as well as worldwide, of admitting that he had been proved wrong, most of all by Americans. Derjaguin stands in shining contrast to Blondlot, of N-ray fame.

Another distinction between polywater and N rays is in the nature of the phenomenon reported. It is clear now that the presumed detection of N rays was due to self-delusion by Blondlot and those who confirmed his work; no such rays can be detected by objective methods. In contrast, water with the properties described by Derjaguin can be prepared; it is the molecular interpretation that is distorted.

On the other hand, both discoveries elicited sociological responses typical of many new inventions. In each case, while some individuals claimed the "discovery" was wrong, others claimed that it was not new but had previously been made by earlier scientists. For polywater, we find two major claims of prior discovery, one citing the American J. L. Shereshefsky, the other the British scientist D. H. Bangham (working in Cairo).

While working on his thesis research (under the guidance of W. Patrick) Shereshefsky had studied the vapor pressure of water in small capillaries. This work was published under his name alone in 1928. Subsequently, as a faculty member at Howard University he extended such investigations in pyrex and quartz capillaries and in 1950 (with a colleague, C. P. Carter) published his results. Some of the statements found in his writings are the following:

The lowering of the vapor pressure of water in cone-shaped capillaries of three to ten microns [millionths of a meter] in radius is found to be greater from 7 to 80 times than the values calculated from the Kelvin equation.

In explanation of the observed abnormal lowering of the vapor pressure one of the authors suggested [in 1928] the possible variation of density and surface tension of the ligand.

It is obvious that the Kelvin equation does not take into account the influence of the capillary wall.

The abnormal lowering of the vapor pressure *may* perhaps *be explained on the basis of a wall effect on the vapor.*

Some of the statements in Bangham's papers in the 1930s were assembled by his son A. D. Bangham and published in 1968.

[P]olymolecular films [of water, etc.] formed [on mica] at saturation have properties quite different from the bulk liquids.

[I]t has been found possible to build up absorbed layers of water and of organic vapors on mica to visible thickness (from supersaturated vapours) and to demonstrate their incongruity with the normal liquids....

Since ... *water adsorbed in coal* (and other carbonaceous solids) *is structurally different from ice and from bulk water,* it is the more important to devise experiments to ascertain its properties.

It is clear from both groups of selections that each investigator felt that water near a surface was subject to forces and influences from the solid container or matrix. Neither Shereshefsky nor Bangham had the courage or the foolhardiness to claim that a change in structure was occurring independent of the neighboring solid. Neither one merits either the credit or the scorn that belongs to Derjaguin.

Novel sources of anxiety, unrecognized at the beginning of this century, are the social, political and environmental consequences of a scientific discovery. Such concerns had reached a crescendo by the 1960s. It will come as no surprise, therefore, that prophets of doom sounded exaggerated alarms almost immediately after poly-water reached the scientific public. Thus in 1969 a letter was sent to *Nature* warning of the perilous risks of research with polywater:

A report on the properties of "anomalous" water appeared recently in *Nature...I need not spell out in detail the consequences if the polymer phase can grow at the expense of normal water under any conditions found in the environment. Polywater may or may not be the secret of Venus's missing water. The polymerization of Earth's water would turn her into a reasonable facsimile of Venus.*

There are examples of phases in other systems which are difficult to nucleate. Once the nuclei are present the phases grow readily, often by mechanisms other than those required to form the nuclei. It is almost a truism that, under conditions where both a stable phase and a metastable phase may form, the metastable phase forms first. In this case the metastable phase would be normal water.

After being convinced of the existence of *polywater, I am not easily persuaded that it is not dangerous. The consequences of being wrong about this matter are so serious that only positive evidence that there is no danger would be acceptable.* Only the existence of natural (ambient) mechanisms which depolymerize the material would prove its safety. Until such mechanisms are known to exist, *I regard the polymer as the most dangerous material on earth.*

Every effort must be made to establish the absolute safety of the material before it is commercially produced. Once the polymer nuclei become dispersed in the soil it will be too late to do anything. Even as I write there are undoubtedly scores of groups preparing polywater.

Scientists everywhere must be alerted to the need for extreme caution in the disposal of polywater. Treat it as the most deadly virus until its safety is established.

Of all people, it was J. D. Bernal (and his co-workers) who felt obliged to respond to this frightening alarum from F. J. Donahoe, of Wilkes College in Pennsylvania:

Dr. Donahoe's unduly alarmist and misleading letter concerning anomalous water has come to our attention. As one of the groups currently trying to sort out the clues surrounding this phenomenon, we feel a reply is called for, especially *considering the alarming newspaper reports to which the letter has given rise.*

Contrary to the data which Dr. Donahoe quotes as fact, remarkably little is still known about the precise properties of the substance, and *it is still not certain that it even exists . . . .*

One of the main reasons for there still being no coherent self-consistent picture of anomalous water is the extreme difficulty of making it in quantities other than microlitres — and there is some suspicion that larger quantities are unstable. *In the laboratory — where extreme care is taken — there is no evidence of its ability to grow at the expense of the normal phase* (with which it is partially miscible). . . *we are sure that not a single worker in the field shares Dr. Donahoe's science fiction worries. . . .One of the greatest difficulties in even accepting the existence of a more stable phase is its apparent absence in nature.* Indeed, this is the most persuasive evidence of its inability to grow at ordinary water's expense, for it has stood the test of billions of years. The classic conditions for its formation — a quartz surface and greater than 95 percent humidity — are very wide-spread in nature, yet no anomalous water has been detected. If it can grow at the expense of ordinary water, we should already be a completely dead planet.

Yet we are not, and totally unlikely to become so from this source. *By all means draw the attention of scientists to the dangers of their work, but make sure it is a real danger before alarming everybody else.*

Alarms such as Donahoe's were grist for the mills of the newspapers. Journalists are always looking for sensational science, and polywater was like sugar water to flies. Water is a substance with which the public is very familiar. Of course in that case it is not news. But "what if" stories can attract attention, particularly those that can raise a few welts on the bodies of "irresponsible, mad scientists." And so inflated versions of Donahoe's concerns appeared in the popular press.

Another journalistic "what if" story is of the science-fiction type and focuses on the enormous potential of a new discovery, to minds unfettered by much knowledge. As an example of this genre we can summarize an item that appeared in the *Wall Street Journal* , that paragon of sound reporting of basic knowledge to the managers of the financial community:

A few years from now living room furniture *may be* made out of water. The antifreeze in cars *may be* water. And overcoats *may be* rainproofed with water. *If* the polywater people are right, the impact on science and industry could be spectacular. Since a polymer of ethylene was first produced in the early 1930s, polyethylene plastic has grown to a $5 billion-a-year industry.

Some researchers say that polywater has an equally rosy future. Soviet scientists assert that the liquid can *probably* be reduced to solid form by heating. *If so,* goods ranging from chairs to piping could be manufactured from a material made literally from water.

Our excursion through the history of water has not ended, however. Fascination with water and "discoveries" of new and unusual properties of water continue. In 1978, five years after even Derjaguin conceded that polywater was a fiasco, the Information Office at the Soviet Embassy issued a release describing still another form of water discovered by two brothers, Igor and Vadim Zelepuklin of the Institute of i.uit-Growing and Vine-Growing in Kazakhstan. This material was claimed to be more active biologically than ordinary water. The release informed us that Soviet scientists have long recognized that water from freshly melted snow can stimulate some biological processes. For example, cut leaves absorb several times more meltwater than either tap water or boiled water. Reminiscent of the guidance of Michurin and Lysenko in Soviet genetics of the middle twentieth century, they also soaked cotton seeds in their "bio-active" water, and found that the cotton plant arising from such seeds produced a greater yield, excelled in physiological characteristics and could be spun into a superior fiber. Equally striking results were obtained with animals. As usual a suitable theoretical explanation was offered: water from melted ice was said to retain in the liquid state some of the molecular order of the crystal; such order was attributed also to water in living cells and was claimed to accentuate enzyme activity.

These special insights are not restricted to scientists in the Soviet Union. Recently, three Frenchmen, J. R. Beaumont, L. Valageas Berger and M.-M. F. E. Frey obtained a patent to cover a new form of water, also prepared from melted snow, with very useful medical applications, particularly in burn therapy. A summary of this document follows:

PROCESS FOR OBTAINING A LIQUID FROM FRESH SNOW FOR BIOLOGICAL TREATMENTS AND PRODUCT COMPOSITIONS USING THIS LIQUID
This invention describes a process for preparing a biologically active liquid by melting fresh powdered snow at a temperature not exceeding 15°C, and, after decantation, subjecting the liquid to an ageing process by storing it in inert containers containing an inert atmosphere (e.g., nitrogen). Care and precision are essential in the collection of the snow in order to preserve the special qualities of the fresh snow. The liquid may be sterilized by ultraviolet rays. After storing for several months it has an odor similar to fresh oysters. The liquid may be formulated into a practical composition by mixing with various constituents such as alcohols, perfumes or creams, and may be converted into an aerosol by using freon under pressure. The product has numerous applications in biological treatments. In superficial burns and sunburn, the product gives immediate relief from pain and leads to decongestion and rehydration of the skin, returning the epidermis to its initial physiological state without any blistering. The product may also be used as a lotion, eau de toilet, etc. It has many applications besides those described here.

One can reasonably expect that discoveries of this type will continue to appear. There will always be individuals with a talent for tailoring irrelevant facts to suit distorted theories.

Allen, L. C., and Kollman, P. (1970). "A Theory of Anomalous Water", *Science,* Volume 167, pp. 1443-1454.

Anonymous (1959). "Svante Arrhenius (1859-1927)", *Endeavour,* Volume 18, pp. 59-60.

Bangham, A. D., and Bangham, D. R. (1968). "Very Long-range Structuring of Liquids, including Water, at Solid Surfaces", *Nature,* Volume 219, pp. 1151-1152.

Beaumont, J. R., Valageas Berger, L., and Frey, M.-M. F. E. (1970). "Process for Obtaining a Liquid from Fresh Snow for Biological Treatments and Product Compositions Using This Liquid", French Patent No. 2,033,769.

Bernal, J. D., and Fowler, R. H. (1933). "A Theory of Water and Ionic Solution, with Particular Reference to Hydrogen and Hydroxyl Ions", *Journal of Chemical Physics,* Volume 1, pp. 515-548.

Bernal, J. D., Barnes, P., Cherry, I. A., and Finney, J. L. (1969). " 'Anomalous' Water", *Nature,* Volume 224, page 393.

Beyerchen, A. D. (1977). *Scientists Under Hitler,* Yale University Press, New Haven, Conn.

Brand, D. (1970). *Wall Street Journal,* Volume 50, Number 197, page 1.

Davis, R. E. (1970). "Polywater Controversy Boils Over", *Chemical and Engineering News,* Volume 48, Number 27, pp. 7-8.

Davis, R. E. (1970). "Polywater in History", *Chemical and Engineering News,* Volume 48, Number 41, page 73 and page 79.

Davis, R. E., Rousseau, D. L. and Board, R. D. (1971). "Polywater's Evidence from Electron Spectroscopy for Chemical Analysis (ESCA) of a Complex Salt Mixture", *Science,* Volume 171, pp. 167-170.

Derjaguin, B. V. (1966). "Effect of Lyophile Surfaces on the Properties of Boundary Liquid Films", *Discussions of the Faraday Society,* Volume 42, pp. 109-119.

Derjaguin, B. V. (1970). "Superdense Water", *Scientific American,* Volume 223, Number 5, pp. 52-71.

Derjaguin, B. V., and Churaev, N. V. (1973). "Nature of 'Anomalous' Water", *Nature,* Volume 244, pp. 430-431.

Donahoe, F. J. (1969). " 'Anomalous' Water", *Nature,* Volume 224, page 198.

Einstein, A. (1949). "Autobiographical Notes" in *Albert Einstein: Philosopher-Scientist,* Vol. VII of *The Library of Living Philosophers,* edited by P. A. Schilpp, George Banta Publishing Co., Menasha, Wis.

Franks, F. (1981). *Polywater,* MIT Press, Cambridge, Mass.

Henderson, L. J. (1913). *The Fitness of the Environment,* Macmillan Co., New York; reprinted in 1970, Peter Smith, Gloucester, Mass.

Hildebrand, J. H. (1970). " 'Polywater' is Hard to Swallow", *Science,* Volume 168, page 1397.

Israel, H., Ruckhaber, E., and Weinmann, R. (1931). *Hundert Autoren Gegen Einstein,* R. Voigtlanders Press, Leipzig.

Lippincott, E. R., Stromberg, R. R., Grant, W. H., and Cessac, G. L. (1969). "Polywater", *Science,* Volume 164, pp. 1482-1487.

Maugh, T. H., II (1978). "Soviet Science: A Wonder Water from Kazakhstan", *Science,* Volume 202, page 414.

Nash, L. K. (1957). In *Harvard Case Histories in Experimental Science,* edited by J. B. Conant, Vol. 2, Case 5, Harvard University Press, Cambridge, Mass.

Rabideau, S. W., and Florin, A. E. (1970). "Anomalous Water: Characterization by Physical Methods", *Science,* Volume 169, pp. 48-52.

Rossotti, H. S. (1971). "Water: How Anomalous Can it Get?", *Journal of Inorganic and Nuclear Chemistry,* Volume 33, pp. 2037-2042.

Rousseau, D. L., and Porto, S. P. S. (1970). "Polywater: Polymer or Artifact?", *Science,* Volume 167, pp. 1715-1719.

Shereshefsky, J. L., and Carter, C. P. (1950). "Liquid-Vapor Equilibrium in Microscopic Capillaries. I. Aqueous System", *Journal of the American Chemical Society,* Volume 72, pp. 3682-3686.

Sun, M. (1981). "At Long Last, Linus Pauling Lands NCI Grant", *Science,* Volume 212, pp. 1126-1127.

V

People Yearn to Believe: Dr. Fox Experiments

There are many ideas widely accepted by different or overlapping publics that are viewed with various degrees of skepticism by critical scientists. Often ingenious experiments have been designed to test or challenge beliefs in such ideas. Even when such experiments have produced results in line with those expected by the skeptical challenger, the findings have not been persuasive in changing a strongly-held viewpoint. Individuals with a deep emotional or intellectual commitment can discover ways to rationalize seemingly contradictory information so that they can retain their principal premises. Let us examine some examples of such confrontations, and the responses thereto.

One experiment that I found intriguing was carried out a few years ago to test the general presumption throughout the academic world that students can tell who is a "good teacher." In general, no attempt is made to clarify in one's mind what features characterize a good teacher. If asked to do so, one's general inclination is to say "Don't ask me to define a good teacher; I can recognize one when I hear one." But can you?

About a decade ago, a team of medical educators, D. H. Naftulin, J. E. Ware, Jr., and F. A. Donnelly, set out to test this presumption. They constructed what they called " a paradigm of educational seduction," which was tested on a selected audience and followed by a written evaluation protocol.

Specifically they trained a professional actor to deliver a lecture, empty of any substance, in a dynamic, charismatic and authoritative manner. The lecturer was chosen to be distinguished in appearance and had conferred on him an appropriate title (and name), Dr. M. L. Fox, and an impressive set of (fictitious) scientific credentials, which were emphasized by the introducer. Dr. Fox, reputed to be an expert in the application of mathematics to human behavior, gave a lecture on that topic as it related to physician education. He had been trained by the authors to use double talk, non sequiturs, unknown terms newly-invented for the occasion and contradictory statements, and to intersperse them with humorous comments, anecdotes, and inane allusions to seemingly related but actually irrelevant subjects. These he incorporated massively not only in his one-hour lecture but also in the thirty-minute question and answer session that followed.

After the lecture and question session each member of the audience assessed the performance by filling out (anonymously) a written evaluation questionnaire, designed to appear totally authentic. The audience included psychiatrists, psychologists, and psychiatric social workers as well as other professional educators not involved with medical health training, in other words, individuals who were not green freshmen. The contents of the questionnaire and the distribution of replies from the audience are listed in Table 1.

A perusal of this Table shows that the response of the audience was overwhelmingly complimentary to the lecturer. (A "No" response to item 1, and a "Yes" answer to items 2–6 were considered favorable). One listener even claimed to have read the speaker's (non-existent) publications.

As in most current collegiate questionnaires for evaluation of faculty teaching, the listener was invited to make specific comments on his general impressions of the lecturer. The similarities in subjective remarks made by the professionals in the audience and those made by undergraduates are striking, both the favorable and unfavorable ones. A sampling of the favorable ones reads as follows:

> Excellent presentation, enjoyed listening. Good flow, seems enthusiastic. Has warm manner. Lively examples. Knowledgeable. Extremely articulate. His relaxed manner of presentation was a large factor in holding my interest. Good analysis of subject that has been personally studied before. Very dramatic presentation. He was certainly captivating.

Such a set of comments, combined with the ratings in Table 1 would have qualified the lecturer for a teaching award in my university.

Table 1. Lecture Evaluation Questionnaire and Responses

Question	Number of Responses	
	Yes	No
1. Did the lecturer dwell upon the obvious?	15	40
2. Did he seem interested in his subject?	53	2
3. Did he use enough examples to clarify his material?	47	8
4. Did he present his material in a well-organized form?	42	13
5. Did he stimulate your thinking?	50	5
6. Did he put his material across in an interesting way?	46	9
7. Have you read any of the speaker's publications?	1	54

Specify any other important characteristics of his presentation.

* Reproduced, with minor modifications, from the article by D. H. Naftulin, J. E. Ware, Jr. and F. A. Donnelly, *Journal of Medical Education,* (1973), vol. 48, pp. 630-635.

Of course in every group of people there are some chronic malcontents and from these came comments such as the following:

> Did not carry it far enough. Left out relevant examples. He misses the last few phrases which I believe would have tied together his ideas. Too intellectual a presentation. Somewhat disorganized. Unorganized and ineffective. Too much gesturing. Frustratingly boring.

However, not one, single individual detected the hoax. And this audience was composed of experienced educators.

This experiment revealed that even experienced educators can feel they have learned something substantive from a lecture with meaningless, irrelevant content, if the lecturer performs in a suitable style and projects a charismatic, authoritative personality.

One can argue that even "the illusion of having learned" serves a motivating function. Indeed, some of Dr. Fox's audience, when informed of the hoax, asserted that the lecture had stimulated them to look into the topic of the talk. There would be general agreement that motivating a student to learn should be one central aim of the teacher. One necessary condition, however, does not provide sufficient criteria to detect a "good teacher." As the Dr. Fox experiment showed, an interesting and exciting lecturer may impart no substantive content. Nevertheless, teaching evaluation questionnaires since the Naftulin-Ware-Donnelly experiment have not been modified to detect "Dr. Foxes."

Let us turn to a different segment of the population and another widely-held belief.

In any year there are numerous reports in the newspapers describing claims by psychics of having solved major crimes. Anecdotal reports of this type have circulated for many decades. Let me describe three examples from those assembled from a variety of sources by Paul Tabori in his book *Crime and the Occult.*

One of the most famous psychics specializing in criminology was W. de Kerler, who practiced at the beginning of this century. One of his exploits is recounted as follows. On a visit to a police chief on the French Côte d'Azur, de Kerler was told of a large theft, the night before, from a prominent local family. The burglar was still unidentified, and de Kerler was challenged to describe him. The psychic proceeded to the scene of the crime, where he discovered a palm print. It contained parts of the "headline," "lifeline" and "heartline." Entering into a trance, de Kerler visualized the criminal and described the man's eyes, hair, precise height, general build and idiosyncratic gait. On the basis of this information, the police chief selected three photographs of individuals that fitted the description given by the psychic; unhesitatingly de Kerler chose one. With that lead, the police quickly apprehended the suspect. A palm print was then taken at the station and found to match perfectly the one found at the scene of the crime.

A more contemporary practitioner of psychic criminology was Marinus Dykshoorn, who claimed to have extra-sensory perception of taste, smell, hearing, touch and sight. Furthermore, by looking at slides with blood smears, he could specify the age, sex and physical characteristics of the donor. Around 1970, when he emigrated to the United States he lived in North Carolina and was purportedly consulted by the police in several murder cases. One of these involved an elderly couple who had been shot and killed in their farmhouse, which was then burned to destroy any evidence. When Dykshoorn visited the ruins, he went into a trance and reenacted the murder scene, revealing that the assailant had been hit on the nose. Furthermore he mimicked the criminal's motions, as he set fire to the dwelling, with such fidelity that the local observers could quickly identify the suspect. Moreover, Dykshoorn in his trance reconstructed the criminal's path of escape and revealed the location of the man's home. With this information a hospital check was made, and it confirmed that the suspect had indeed entered there with an injured nose at the time of the crime. The criminal was apprehended.

Finally we cite Alex Tanous of Maine, who purportedly was instrumental, in 1972, in solving a particularly brutal murder of an eight-year-old child. Standing in the victim's apartment house, the psychic revealed that the body was somewhere in that building, under "something." Furthermore, on the following day he produced a sketch that he had drawn of the murderer. The police chief thought this drawing matched in detail that of a photograph of one of the suspects. A search of the suspect's apartment disclosed the victim's body under a bed.

These and many other anecdotes that appear regularly in the newspapers always describe claims made by psychics post hoc. Their success with estrasensory perceptions are always disclosed after the police have identified or apprehended the criminal.

Obviously such "sensitive" individuals could be of enormous service to public safety agencies. What we need, however, is some convincing demonstration of their endowment with extrasensory perception *before* the crime is solved.

A clever "Dr. Fox" experiment was designed for this purpose by a group of scientists (M. Reiser, L. Ludwig, S. Saxe and C. Wagner) with connections to the Los Angeles Police Department. Southern California probably has more than its share of gifted individuals, so it was not difficult to assemble a dozen psychics for the project. The project was initiated shortly after a series of rape-murders had taken place in Los Angeles, and had been widely publicized, and had been christened with a suitably lurid name, the "Hillside Strangler" cases. The publicity had attracted many offers of assistance from individuals claiming to possess extrasensory perception. Dr. Reiser and his colleagues designed a study to see if such gifted individuals could aid in the apprehension of suspects.

Four crimes were selected, none related to the Hillside Strangler cases. In the best scientific tradition, a double-blind plan was devised: neither the psychics nor the administrator of the test (a psychologist) had any prior acquaintance with the available clues or the nature of the crime. The clues for each crime were placed in sealed, numbered envelopes. Two of the four crimes had been solved, two were still unsolved.

Among the dozen psychics, eight normally earned fees for their services, and four were non-professionals who had reputations for being among the most capable. In separate sessions, each psychic was first shown the sealed envelopes, for one specific case, and asked to educe any relevant information. Then this individual was asked to open the envelopes, examine the contents and draw out any additional useful facts. Responses were recorded on tape.

Only about half the information elicited through the dozen psychics was verifiable in nature. Many of the psychics believed they were being shown evidence from the "Hillside Strangler" case, but the experimenters chose crimes totally unrelated to that in order to circumvent suggestive speculation from the newspapers. Some individuals were extremely talkative but provided little substantive information. One representative response ran as follows:

> I get a man, black. I hear screaming, screaming. I'm running up stairs and down. My head . . . someone bounces my head on the wall or floor. I see trees — a park? In the city, but green. Did this person live there? What does the number "2" mean? I get a bad, bloody taste in my mouth. The names "John" or "Joseph" or something like that. I am running on the street like a crazy. This is a *very* serious crime. I can't hold the envelope in my hand.

From such repsonses the experimenters filtered out data that could be placed in one of five broad categories for which there could be definitive information in the crime report:

(1) Nature of crime
(2) Victim
(3) Suspect
(4) Physical descriptions
(5) Location of crime.

Let us examine the detailed responses in two of the crimes.

The facts in the first case are as follows. The victim was a white female, thirty years old, killed in the course of a robbery. At the time of the experiment, the suspect had not been apprehended. The clue in the envelope was a cigarette lighter. Table 2 lists each item for which the individual psychic provided correct specific statements. Of the twenty-one different classes of items listed in the Table headings, no more than four were identified correctly by any psychic.

Table 2. Correct Responses from Psychics in Crime No. 1

Psychic Number	Victim's Name	Crime Location	Address	Sex	Descent	Age	Hair	Eyes	Height	Weight	Build	Complexion	Clothing Worn	Cause of Death	Occupation	Date Occurred	Time Occurred	Property Taken	Modus Operandi	Weapon	Crime	TOTALS
1																						0
2				•	•				•												•	4
3																						0
4				•																		1
5-A*					•													•				2
6-A					•																•	2
7-A				•	•	•																3
8																				•	•	2
9-A				•	•		•														•	4
10				•			•															2
11				•																		1
12																			•	•	•	3
TOTALS				6	5	1	2		1									1	1	2	5	

* A = Nonprofessional.

With two of the classes, sex and descent, the correct response rate was about fifty percent, in other words, no better than chance. In sum, nothing of any use in recognizing the criminal was provided.

In the second crime the victim was an unusual one, an eighty-nine-year-old man, a church historian, again one killed during a robbery. In this case, however, the suspect had been apprehended. The clues shown to the psychics were a lens from eyeglasses and a pair of tassled, loafer shoes. The correct responses from the psychics are listed in Table 3. Of the thirty-three different classes of items listed, three was the maximum number correctly specified by any individual. Six did correctly specify the nature of the crime and seven chose the correct sex. Essentially nothing about the characteristics of the victim was correctly educed, but one psychic did insist that a church was pertinent to the crime.

Table 3. Correct Responses from Psychics in Crime No. 2

Psychic Number	1	2	3	4	5-A*	6-A	7-A	8	9-A	10	11	12	TOTALS
Victim's Name													
Crime Location													
Address													
Sex							•						1
Descent													
Age													
Hair													
Eyes													
Height													
Weight													
Build													
Complexion													
Clothing Worn													
Cause of Death													
Occupation													
Date Occurred													
Time Occurred													
Property Taken													
Modus Operandi													
Weapon													
Suspect's Name													
Crime	•			•		•	•	•				•	6
Address													
Sex	•	•		•	•				•	•		•	7
Descent		•		•									2
Age													
Hair													
Eyes													
Weight													
Height		•							•				2
Physical Oddities													
Occupation													
Clothing Worn													
TOTALS	2	3	0	3	1	1	2	1	2	1	0	2	

*A = Nonprofessional.

Returning to Crime No. 1, the experimenters made one more interesting analysis. Since the suspect was not in hand, a comparison was made among specifications given by the psychics to see how congruent they were. Ten psychics said the criminal was male; two said he was a white, two a black. Considering the proportion of females among perpetrators of violent crimes, one can conclude that the agreement among psychics is about the same as that from guesswork.

Does this set of experiments invalidate the claims of psychics, at least in regard to criminal cases? Hardly. Experts in the field of extrasensory perception point out that many factors affect the responses of a psychic, and therefore a statistical analysis of correct hits is not appropriate. We are told that a psychic must be properly motivated to respond correctly. Hence extrasensory effectiveness may vary from day to day, or under different circumstances. A psychic may be sensitive in one case but totally off the mark in another. Or in any individual case, correct information may be educed for some aspects but not others. (Thus in Crime 2, one psychic did state that a church was somehow relevant, although this statement was of no help in identifying or apprehending the criminal.) Even more unsettling is the possibility that the psychic's perception about a case is correct, but that the information obtained has been assigned to an alternative case, the wrong one. With these rationalizations, no experiment could disprove the premise that extrasensory perception is effective in police detective work. Thus one's position on this issue after the "Dr. Fox" experiment of Reiser and his colleagues would probably be identical with that held initially. The Los Angeles experimenters tactfully concluded that "the usefulness of psychics as an aid in criminal investigation has not been validated," and recommended that "further research in this area would be desirable." It would be interesting to learn Senator Proxmire's response to this recommendation, if it were to include a request for federal funding.

Religious relics provide a particularly fertile field for skeptics to carry out "Dr. Fox" experiments. However, in this area deep and pervasive feelings can be bruised easily and arguments can become very emotional. Again it is doubtful that any one's conceptual commitment is changed in consequence of these arguments.

Although the Shroud of Turin is currently the most fascinating example of this social phenomenon, a crucial test from a skeptic's viewpoint (i.e., carbon-14 dating) cannot be carried out until church authorities permit it. Consequently, heated exchanges revolve around secondary aspects. Instead I should like to examine a subject, purported remnants of Noah's Ark, that is not inherently an object of such deep, sacred veneration.

According to the biblical account in Genesis VIII, Noah's Ark came to rest on the top of the mountains (note the plural) of Ararat. What is currently called Mount Ararat, is now a snow-capped peak rising to almost 17,000 feet above sea level located at the northeastern end of Turkey, close to the borders of the Soviet Union and Iran. Even before present-day turmoil in the area, the region was not easily accessible, nor one in which the Turkish military welcomed visitors.

Anecdotal stories testifying to the presence of the Ark on Mount Ararat go back centuries. Even as early as the first century A.D., Josephus, writing in Rome, reported that remains of the Ark were being shown by Armenian inhabitants of the Ararat region. Within the past hundred years there have been a number of claims of sightings of an ancient wooden ship high up on the mountain, partly buried in snow. There is indeed a spur of rock, not far from the top, that an active imagination could visualize as the bow of a ship, but it is made of stone, not wood. Various teams of mountain climbers, or individuals, claimed to have ascended to the top of the mountain during the nineteenth century and to have seen remnants of the Ark. None brought back any evidence that could be inspected by interested parties. Finally, however, in the 1960s, a French industrialist, Fernand Navarra, found some wood above the timberline and returned to ground level with it. His conclusion that he had been successful in finding remains of the Ark is described in the book he published in 1974, *Noah's Ark: I Touched It.*

Navarra's qualifications for the search are described in his narrative: when he was a child, his acquaintances claimed that he had an instinct for finding lost objects, for he was successful in helping them track down missing things. Subsequently, he served with the French Army in the Middle East where he became an expert climber and where he heard anecdotal stories about the Ark. Some time after World War II, he undertook to climb Ararat (Figure 1). After several expeditions he succeeded in finding pieces of "obviously hand-hewn" wood at the bottom of a crevasse in ice at 14,000 feet and he carried these down and back to France. Figure 2 shows views into the crevasse; Figure 3 is a reproduction of Navarra's extrapolation of the information in Figure 2.

Figure 1. Photograph of Mount Ararat taken by F. Navarra during his expeditions to find Noah's Ark. (Reproduced with permission of Mr. Navarra.)

One of the several photographs of pieces of the recovered wood is shown in Figure 4. According to Navarra these pieces had been hand-hewn and squared. To preserve them he created a museum exhibit in France, which includes a model of the ark built by Navarra on the basis of his researches (Figure 5).

What is the evidence that the wood is a remnant of Noah's Ark? Well, as Navarra said:

> What else could it be?...I found this wood on Ararat, the traditional stopping place of the Ark. And this wood, besides being visibly squared, cannot be a plain tree trunk. What kind of construction took place on Mount Ararat at nearly 14,000 feet, almost 5,000 years ago?

Navarra also had samples of the wood examined by: (1) the National Museum of Natural History in Paris, which concluded that the sample was oak; (2) the Spanish Forestry Institute of Research and Experiments, which concluded that (a) the sample was oak, and (b) the density of 1.100 (normal wood runs between 0.800 and 0.850)

Figure 2. Crevasses in ice on Mount Ararat. From this cleft Navarra recovered pieces of wood that he claimed to be remnants from Noah's Ark. (Reproduced with permission of Mr. Navarra.)

and the color permit one to "suppose that the age of the wood... [is] around five thousand years"; (3) the Director of the Prehistoric Institute of the University of Bordeaux who reported that (a) the wood was oak, (b) its density was 0.938 (contrast with 1.100 measured by (2)) and (c) "the advanced state of ligninization... [assigns the sample to] a period dating to a remote antiquity."

To further support his claim, Navarra included in his book a letter from Mr. Carlton Yerex (responding to one from Navarra) stating "I am happy to read [where? in Navarra's letter?] that the exact age of the wood has been dated by the radiocarbon process... [and that it is] 4484 years old." I am unable to discover in Yerex's letter or Navarra's book any statement as to who carried out the radiocarbon dating. In any event Mr. Yerex then proceeded to say that 4484 years

Figure 3. Navarra's view of the location of the crevasse from which wood was recovered. (Reproduced with permission of Mr. Navarra.)

Figure 4. Photograph of one of the pieces of wood found by Navarra high on Mount Ararat. (Reproduced with permission of Mr. Navarra.)

Figure 5. Model of Noah's Ark constructed by F. Navarra on the basis of his researches. (Reproduced with permission of Mr. Navarra.)

is *exactly* what one would expect from biblical chronology, which he demonstrates as follows:

Date of the flood (B. C.)	2472
Date in which the Ark was found	1955
Number of years from the flood to 1955	4427
Number of years wood was cut before the flood	57
Total age of the wood when it was examined	4484

Since all of this is included in Navarra's book, I presume it is meant to substantiate his claim to have recovered a remnant of Noah's Ark.

How and when did Dr. Fox enter this story? In a sense I am unsure, because there is room for different views as to who is Dr. Fox in this narrative. In any event, some pertinent facts are the following. Navarra's book appeared in 1974. However, as early as 1965, a report on carbon-14 dating of Navarra's wood was published, from a highly reputable source, the National Physical Laboratory in Great Britain (the equivalent of the National Bureau of Standards in the United States). This estimate has been confirmed subsequently by several other groups of scientists. Their results are summarized in Table 4.

Seven samples of the wood recovered from the crevasse on Mount Ararat have been dated by radiocarbon determinations in six different laboratories. Five of the determinations place an age of 1200–1300 years on the wood. The other two assign a somewhat older age. Nevertheless all agree that the wood did not exist on earth before the first century A.D. Most of the assays place the birth of the wood in the seventh to the eighth century A.D. From the genealogy listed in Genesis V, "the book of the generations of Adam," we can calculate that Noah's ark was constructed 1660 years after the creation of Adam. Using Bishop Ussher's assignment of the year 4004 B.C. as that of the Creation, we conclude that wood from Noah's ark ought to be about 4300 years old, not the 1300 years found from radiocarbon dating.

Does this settle the issue of the authenticity of the relic from Noah's ark? Hardly. A believer will conclude that the radiocarbon clocks must be off. This view is clearly expressed by individuals such as Dr. Harold Slusher, an assistant professor of physics in Texas (El Paso) and a member of the Institute for Creation Research. Dr. Slusher lists many possible reasons why C-14 dating may be unreliable: cosmic-ray bombardment *may have been* different in the past; the earth's magnetic field is (*purported to be*) decreasing rapidly with time; a vapor canopy *may have* covered the earth before the Noachian Flood; the content of nitrogen atoms in the atmosphere from which radiocarbon is obtained by cosmic-ray bombardment *may have been* different in the past; the radiocarbon clocks *may have been* going faster in the past, etc.

In general, these responses to dating discrepancies are ad hoc assumptions; occasional references to some supposed evidence are mostly to publications that have not undergone scrutiny by experts in the physics of geochronology. A rationalist may feel obliged to

Table 4.* Historical Place in Time of Wood from Mount Ararat

Laboratory Number	Radiocarbon Age	Historical Place
NPL-61	1190 ± 90	AD 770–790
UCR-553	1210 ± 90	AD 730–760
UCLA-1607	1230 ± 60	AD 730
P-1620	1320 ± 50	AD 640
GX-1667	1350 ± 95	AD 620–640
GX-1668	1690 ± 120	AD 270
I-(?)	exact number not available	less than 2,000 years old

*Adapted from table in article by R. E. Taylor and R. Berger, *Antiquity,* LIV, pp. 34-36, March 1980.

examine each of these criticisms in detail, but if he does he will become mired in a morass. For example, in connection with the specific case of wood from Noah's Ark, it has been suggested that the C-14 content was influenced by the altitude at which the samples lay for thousands of years; purportedly solar protons could collide with N-14 in situ in the wood to produce additional radiocarbon beyond that incorporated by the living tree from which the wood sample came. After substantial investigation, a number of scientists showed this factor could not be significant; for example, it has been shown that samples of high-altitude bristlecone from the White Mountains in the United States and low-altitude European oak of the same botanical age have the same C-14 activities. Nevertheless, one could argue that circumstances are different in Asia. And such counter-exchanges can go on almost indefinitely. The irony of this scenario is that the proponent of a claim does not feel obligated to prove it but throws the burden — disproof — on his opponents. The ultimate position that a strong believer can take is that when the Creator brought the universe into existence some six millennia ago he built in the entire range of C-14 contents that we now measure in archaeological and paleontological artifacts. This statement is not subject to scientific discrediting.

Another demand of strong adherents to a belief is that their critics provide alternative explanations of the items of "evidence" they have presented. This presumption is based on a fallacious understanding of the character of scientific judgment. One does not attempt to fit every piece of alleged information into the general assessment of an idea. As Hudson Hoagland once said (in another connection, i.e., psychic phenomena and UFOs)

There will [always] be cases which remain unexplained because of lack of data, lack of repeatability, false reporting, wishful thinking, deluded observers, rumors, lies, frauds....Unexplained cases are simply unexplained. They can never constitute evidence for any hypothesis....The basic difficulty...is that it is impossible for science ever to prove a universal negative.

Thus Navarra asked, if the wood he discovered near the top of Mount Ararat is not from Noah's Ark, "what else could it be?" Actually, once the relic had been dated by radiocarbon analysis, several individuals pointed out that there has been a long tradition of erection of symbols of religious veneration near places of sacred status. The construction of shrines to commemorate biblical events has been especially strong among Armenian and Byzantine Christian groups. Thus it is very reasonable to presume that the wood recovered by Navarra is from the remains of a cenotaph, cross or shrine constructed near the top of Mount Ararat in the seventh to the eighth century. In fact, in his book, Navarra showed a photograph of crosses carved into the rocks of the mountain near areas where hermits lived in the caves.

The argument over Navarra's wood is only one example of the continuous quarrel between Creationists and Rationalists over a variety of issues, of which evolution is the current major focus of contention. Creationists propose that such disagreements should be settled by debate, in public forums. But there are some issues that the public may not be qualified to judge. From experience with nonscience college students, I know that the majority predict that if a silver dollar and a silver dime are simultaneously dropped from the top of the Sears Tower the former will hit the ground first. An even larger fraction when asked whether a two-inch long line or a one-inch line has more points will vote in favor of the former; the remainder exercise the democratic privilege of not voting. The general public would overwhelmingly endorse the view of the majority of nonscience students, both groups correctly pointing out that their stand was the common-sense one. And they are right; it *is* what one would expect from common sense. Likewise common sense would discard the principles of relativity and of wave mechanics. Unfortunately for common sense, science does not always agree with it, especially in modern times when the range of scientific experience has far outrun that to which we are exposed in common, man-scale experience.

People yearn to believe (Figure 6). In many circumstances, people even prefer to be deceived — *populus vult decipi* said the ancient Romans. It is not an obligation of scientists or humanists to argue with individuals in such circumstances. Rather we should have compassion for them, for they have placed such unnecessarily rigid constraints on their outlooks. As rational human beings, our goal should be (to paraphrase Saul Bellow) to recognize "the bottles into which *we* have been processed" and to search for directions that will facilitate release from the conceptual constraints in our own views of the world within and around us.

Cooler
Cloudy with rain likely; high in the 70s. Details on Page 72.

★★★★★
Metro
Final

Sun-Times

Chicago, Tuesday, August 24, 1982

25¢ city and suburbs; 40¢ elsewhere

Solid proof of Noah's ark reported found

ANKARA, Turkey (UPI)— A Turkish-American member of the expedition led by former astronaut James Irwin Monday said it had found evidence that the biblical Noah's ark came to rest on Mt. Ararat.

Yucel Donmez, 36, a resident of Chicago, said, "The expedition has revealed solid evidence that the ark is lying up there."

He said a second expedition to the summit of the 16,946-foot peak, taking along better technical equipment and climbing from the northwest side, is needed to produce the evidence.

Irwin's expedition up the north side was cut short by several days when the former astronaut was hurt in a 100-foot fall from an ice ridge while approaching the summit Thursday.

Irwin was flown Saturday to a military hospital in Erzurum. 445 miles east of Ankara. Doctors said he will be released Wednesday.

Irwin said his expedition on the mountain's north side failed to find any evidence of the ark. After he was evacuated from the peak, however, other expedition members remained behind.

Donmez, who scaled Mt. Ararat for the 63rd time last week, and the 10 other members of the team returned to Erzurum Sunday night after a nine-hour bus trip from a military camp near Turkey's highest mountain.

If permission is given, Donmez said, an aerial search of the mountain will be conducted Tuesday.

Meanwhile, U.S. Embassy officials in Ankara said other expedition teams have applied for Turkish government approval to scale Mt. Ararat.

Since 1972, foreigners generally have been prohibited from climbing the peak overlooking the Soviet border for security reasons.

Irwin's expedition, sponsored by a fundamentalist Christian group in Colorado Springs, Colo., was granted permission to scale the peak by Gen. Kenan Evren, Turkey's military leader.

Nation/world

★ ★ Chicago Tribune; Sunday, August 26, 1984 Section 1 3

Explorers find boat formation in Noah's territory

From Chicago Tribune wires

ANKARA, Turkey—U.S. explorers, including former astronaut James Irwin, have found a boat-shaped formation on Mt. Ararat that they believe is the site of Noah's ark, the group's leader said Saturday.

Marvin Steffins, president of International Expeditions of Monroe, La., told reporters that his group found the site Thursday 3,306 feet up the southern slope of the mountain in eastern Turkey.

"We cannot say that this is Noah's ark," Steffins said, "but we believe we have found the site of it."

His expedition found a boat-shaped impression with measurements fitting the biblical description of the ark in the sixth chapter of the Book of Genesis, he said

"BUT I MUST stress that we don't claim to have found the ark of Noah itself," he said, "though the measurements are similar to those in the Bible." The boat-shaped formation is about 450 feet long and 80 feet wide.

The Bible says that after the great flood, the ark carrying a male and female of every species came to rest on "the Mountains of Ararat."

Steffins showed reporters a large bag of samples he brought from the site and said they will be analyzed in the United States.

The samples were oxidized small pieces of flat rocks and wood chips with small packages of sand or soil.

Steffins said the ark could have been almost destroyed by the elements after so many centuries, but

300 cubits long, 50 cubits wide and 30 cubits high. A cubit, an ancient form of measurement, is believed equal to 18 to 22 inches. That would make the ark at least 450 feet long.

MT. ARARAT, called Mt. Agri in modern Turkey, is 17 miles from the Soviet border, a sensitive military area. Turkey lifted a 10-year ban on exploration there in 1982.

Steffins' expedition was one of three U.S. groups that scaled the 16,946-foot mountain this year.

Irwin and his High Flight Society expedition, along with a group of explorers from the U.S. Institute for Creation Research, were continuing their work on the northern slopes, Steffins said.

He said Irwin, who walked on the

he added, "There is enough there to figure out the proper sizes of the instrument, and the archeologists will be able to measure it."

"I AM NOT AN archeologist nor a geologist, so we asked Turkish authorities to grant a permit for continuing archeological work by competent personnel," Steffins said.

Moon 17 years ago, was present at the discovery. So was Steffins' wife, Marjorie, their daughter, Marianne; Louis McCollum of Danville, Ill.; Tim Brentley, a U.S. missionary working in Greece; and Bulent Atalay of Turkey.

Last year Irwin's expedition scaled the peak of Ararat but was forced to cut short its work after the astronaut fell off a cliff and injured his leg.

"We searched this time on the southern side of the mountain, although the general belief was that it was in the north," Steffins said. He said his expedition went to work on the basis of evidence provided by Ron Wyatt, a U.S. explorer who scaled Ararat in 1977.

Figure 6. Announcements of two recent rediscoveries of Noah's Ark on Mount Ararat. It will be of interest to learn how close these finds are to the location of Mr. Navarra's. (Reproduced courtesy of UPI and of Chicago Tribune).

Hammond, A. and Margulis, L. (1981). "Farewell to Newton, Einstein, Darwin...", *Science 81*, December issue, pp. 55-57.

Hoaglund, H. (1969). "Beings from Outer Space — Corporeal and Spiritual", *Science*, Volume 163, page 625.

Klotz, I. M. (1982). "Why Not Teach Creationism in the Schools?", *Bioscience*, Volume 32, pp. 334-335.

Naftulin, D. H., Ware, J. E., Jr. and Donnelly, F. A. (1973). "The Doctor Fox Lecture: A Paradigm of Educational Seduction", *Journal of Medical Education*, Volume 48, pp. 630-635.

Navarra, F. (1974). *Noah's Ark: I Touched It*, Logos International, Plainfield, New Jersey.

Reiser, M., Ludwig, L., Saxe, S. and Wagner, C. (1979). "An Evaluation of the Use of Psychics in the Investigation of Major Crimes", *Journal of Police Science and Administration*, Volume 7, pp. 18-25.

Slusher, H. S. (1981). *Critique of Radiometric Dating*, ICR Technical Monograph No. 2, Second Edition, Institute for Creation Research, San Diego, CA.

Tabori, P. (1974). *Crime and the Occult*, Taplinger Publishing Co., New York.

Taylor, R. E., and Berger, R. (1980). "The Date of 'Noah's Ark'", *Antiquity*, Volume 54, March issue, pp. 34-36.

Van den Bergh, S. (1981). "Size and Age of the Universe", *Science*, Volume 213, pp. 825-830.

Epilogue

"...walk humbly with your Lord."
Micah 6:8.

Acknowledgments

The gestation period for this book was almost two decades. During those years I profited from conversations and correspondence with many individuals. I am particularly indebted to the following:

A. D. Bangham (Cambridge)
Laurie Brown (Northwestern University)
Ralph Burton (North Carolina State University)
Peter Farago (Chemical Society, London)
B. Raymond Fink (University of Washington)
John Finney (Birkbeck College)
David Hargreave (University of New Hampshire)
Frank Herbstein (Technion)
David Joravsky (Northwestern University)
Mark Kac (Rockefeller University)
Joseph Katz (Argonne Laboratory)
Laszlo Lorand (Northwestern University)
E. Georges Merinfeld (Dalhousie University)
Emery Meschter (Kennett Square, Pennsylvania)
Fernand Navarra (Bordeaux)
Franklin Offner (Northwestern University)
Donald Peters (Office of Naval Research)
Curtis Singleterry (Naval Research Laboratory)
R. E. Taylor (University of California, Riverside)
Albert Wolfson (Northwestern University)

The source of the photographs and charts that are not mine are indicated in the legends to the figures. I wish to express my appreciation to the following for their assistance in obtaining the materials:

J. Colomb-Gérard (Paris)
Permanent Secretary, Academie des Sciences (France)
Louise Contis (Swedish Embassy)

Dr. Nils Starfelt (Swedish Embassy)
Wilhelm Odelberg (Stockholm University Library)
Samuel Danishefsky (Yale University)
Ann Bujalski (Yale University)
Bryce Crawford, Jr. (National Academy of Sciences USA)
Victoria Crawford (National Academy of Sciences USA)
James Barsky (Academic Press)
Betty Schlossberg (Harcourt Brace Jovanovich)
Julia Morgan (Johns Hopkins University)
Robert Kieckhefer (UPI/Chicago)
Beverly Fleury (Tribune Media Services)

I owe a special debt of gratitude to my long-time friend and scientific colleague David Volman with whom I have had many rambling and searching discussions over a period of four decades. In addition he initiated the final stimulus for converting my lectures to written form by inviting me to give a series of "public" talks at the University of California, Davis. These became the basis of the essays in this volume. Special thanks are also due to my son Edward Klotz who helped extensively in the foreign translations and who, at my lectures, always laughed heartily at the appropriate points.

Evanston, IL   Irving M. Klotz